EXPERIMENTAL METHODS

IN

FOOD ENGINEERING

S.S.H. Rizvi
Department of Food Science
Cornell University
Ithaca, NY

G.S. Mittal
School of Engineering
University of Guelph
Guelph, CANADA

CBS Publishers & Distributors Pvt. Ltd.

New Delhi • Bengaluru • Chennai • Kochi • Kolkata • Mumbai
Hyderabad • Uttarakhand • Nagpur • Patna • Pune • Jharkhand

CBS Pubs ISBN: 81-239-0528-9
Chapman ISBN: 0-442-00886-4

First Indian Edition: 1997
Reprint: 2004

Copyright © 1992 by Chapman & Hall Inc., New York

This edition has been published in India by arrangement with Chapman & Hall Inc., New York

All rights reserved. No part of this book may be reproduced or transmitted in any form or by any means, electronic or mechanical, including photocopying, recording, or any information storage and retrieval system without permission, in writing, from the publisher.

Published by **Satish Kumar Jain** and produced by **Varun Jain** for
CBS Publishers & Distributors Pvt. Ltd.,
4819/XI Prahlad Street, 24 Ansari Road, Daryaganj, New Delhi - 110002
delhi@cbspd.com, cbspubs@airtelmail.in • www.cbspd.com
Ph.: 23289259, 23266861, 23266867 • Fax: 011-23243014

Corporate Office: 204 FIE, Industrial Area, Patparganj, Delhi - 110 092
Ph: 49344934 • Fax: 011-49344935
E-mail: publishing@cbspd.com • publicity@cbspd.com

Branches:
- ***Bengaluru:*** 2975, 17th Cross, K.R. Road, Bansankari 2nd Stage, Bengaluru - 70 • Ph: +91-80-26771678/79 • Fax: +91-80-26771680 E-mail: cbsbng@gmail.com, bangalore@cbspd.com
- ***Chennai:*** No. 7, Subbaraya Street, Shenoy Nagar, Chennai - 600030 Ph: +91-44-26681266, 26680620 • Fax: +91-44-42032115 E-mail: chennai@cbspd.com
- ***Kochi:*** Ashana House, 39/1904, A.M. Thomas Road, Valanjambalam, Ernakulum, Kochi • Ph: +91-484-4059061-65 Fax: +91-484-4059065 • E-mail: cochin@cbspd.com
- ***Kolkata:*** 6-B, Ground Floor, Rameshwar Shaw Road, Kolkata - 700014 Ph: +91-33-22891126/7/8 • E-mail: kolkata@cbspd.com
- ***Mumbai:*** 83-C, Dr. E. Moses Road, Worli, Mumbai - 400018 Ph: +91-9833017933, 022-24902340/41 • E-mail: mumbai@cbspd.com

Representatives:
- Hyderabad: 0-9885175004
- Patna: 0-9334159340
- Jharkhand: 0-9811541605
- Nagpur: 0-9021734563
- Pune: 0-9623451994
- Uttarakhand: 0-9716462459

Printed at:
J.S. Offset Printers, Delhi (India)

EXPERIMENTAL METHODS

IN

FOOD ENGINEERING

CONTENTS

PREFACE

Some forms of the basic unit operations of food processing and preservation have been practiced by Homo sapiens from well before the time of recorded history. Today, it is indeed difficult to visualize a food processing operation in which one or more of the phenomena of transport processes, thermodynamics, reaction kinetics, and other related principles do not apply. Attempts to analyze them quantitatively are often confronted with formidable challenges due to the limited availability of data on physical and engineering properties of food and food components. This lack of technical information is counterbalanced by vast experience based on empirical techniques. Developments of the scientific approach in quantification of physical and biological factors for application in design and analysis of equipment and processes are of recent origin. The complexity of the food systems also necessitates experimental determination of properties of interest for a meaningful analysis. Although most engineering analyses are rational in concept, their application is often empirical, based on experimentally determined factors emphasizing the need for a knowledge of important experimental methods.

A young generation of food scientists is expected to have a thorough and better understanding of engineering principles in order to become professionals of full stature. The engineering background of direct interest to food processing includes the theory of equilibrium (or thermodynamic) and nonequilibrium (or transport) processes. However, most food science curricula require only one course in food engineering. More often than not, such a course becomes a survey of selected engineering principles. It would indeed be desirable to offer at least two food engineering and two food processing unit operations courses--each a semester long--to food science undergraduates. These courses are expected to include a three-hour laboratory period each week.

A number of texts on food engineering have been published over the last decade, dealing with the principles and practices of theoretical analysis of food processing operations. Although laboratory measurements have become more crucial and sophisticated, no text is currently available that deals with the basic experimental methods in food engineering. It is to fill this gap that this book has been written. The principal objective in doing this has been to enhance the teachability of food engineering principles to food science majors. The material included in the book is based on food engineering laboratory courses that we have offered at our respective universities and all laboratory exercises have therefore been class tested. A list of selected food engineering books for further study and reference is given in the appendix.

The book is organized to permit its use as a text in a food engineering laboratory course. It would be useful to cover the background material in lecture sessions concurrently. For a good understanding of the material presented a knowledge of elementary calculus and applied physics is required. The entire text can be covered in a two-semester, three-hour course. Selectively, but adequately, the book should also serve the need of a one-semester course. The material should also be of value to practicing scientists and engineers.

It is our experience that most students walk into a laboratory class without having read the exercise they are supposed to do. To overcome this, we have included some prelab questions in the text of each exercise that students must answer on the prelab answer sheet prior to initiating the laboratory exercise. This serves as their "ticket" to doing the exercises and thus obligates them to go over the text material in advance. We have found this arrangement quite satisfactory.

We are gratefully indebted to Professor J. L. Blaisdell of The Ohio State University under whom we studied food engineering and learned the importance of experimental measurements. In addition, we acknowledge the considerable help we have received from many colleagues and students, including Steven Mulvaney, Igbal Ali, Greg Ziegler, and Ming Zhang. The help received from several of our department secretaries is also thankfully acknowledged.

Any errors in fact, omissions, and misprints are our responsibility. We will appreciate having them brought to our attention, along with suggestions for enhancing the usefulness of the text.

LABORATORY REPORT WRITING

INTRODUCTION

The content and presentation of a laboratory report may be organized in many forms. It is indeed impossible to establish a format that would suffice the needs for various types of reports. However, a good and effective report has certain essential elements that should be followed. The author has the ultimate choice to make the report most valuable. The report should aim at clarity and precision in presentation. Omit the use of personal pronouns "I", "we," and "you". It is highly recommended that reports should be prepared using only standard size papers, not spiral tear-outs. Reports should be legible and neat. Remember, a well-presented report makes a good impression on the reader.

ORGANIZATION

Described as follows are some fundamental aspects of a good report. The organization may be modified by the instructor for certain types of reports. When in doubt, prepare an outline of your report and discuss it with the instructor.

1. *Cover*: All reports should be submitted in a report cover, preferably transparent.

2. *Title Page*: The first page of the report should include the following items:

   Title of report
   Author
   Name and address of organization where work was performed
   Date

   Additional items such as course number, laboratory exercise number, date of experiment, group members names and the like may be added by the instructor for classroom reports. A title page sample for such use has been included for this purpose.

3. *Summary*: The purpose of the summary is to succinctly provide the essential information to a reader. It must contain a brief statement of the problem and focus on the major results, conclusions, and recommendations. Vague and general statements should be avoided.

4. *Introduction*: Introductory material pertinent to the experimental work should be presented so that readers will follow intelligently the scope of the problems and the proposed experimental work. Where necessary, introduce the prior knowledge and literature with proper reference to the source. The report should conclude with statements on specific objectives to be performed. Even though these objectives are outlined in each

of the laboratory exercises, try to improve on them in your own words. Also, include a statement about how the laboratory exercise relates to food science and/or the food industry.

5. *Materials and Methods*: This part includes a list of equipment and the procedure used in carrying out the work. Often, a schematic diagram of the experimental setup is extremely helpful. Remember, a good diagram is worth a thousand words. The information presented should have enough detail so that other will be able to read it and repeat the experiment in exact detail.

6. *Results and Discussion*: In this section are recorded important data and observations of the experimental work along with their critical discussion. Also included in the written text are illustrative tables, charts, and/or graphs. The original data, if numerous, are included in the Appendix. It should be recognized that graphs, in particular, should "stand on their own" with the ultimate goal that the reader can understand each one without referring to the other. However, no figure or graph should go unmentioned in the written text. A brief summary of the types of calculations performed should be included and detailed calculations are put in the appendix. Figures (i.e., graphs, diagrams) should be numbered consecutively, given complete titles and legends, and the titles should be located below the figures. Tables should be numbered consecutively (independent of figures), given complete titles, and the titles should be located above the table. The results and discussion are so arranged as to indicate relationships and provide answers to questions posed in the list of laboratory goals. From this part emerge justifications for all conclusions about the investigation undertaken. It is, therefore, very important to point out assumptions, sources of error, actual versus expected results, trends, and so on.

7. *Conclusions*: This section should specify logical or warranted conclusions and inferences. It should indicate what has been learned from the work, emphasizing significance and value of the results. When there are a number of conclusions, they should each be listed.

8. *References*: List all reference materials used in preparing the report in the format either used by the *Journal of Food Science* or stated by the instructor. The former follows the author(s) and year format shown as follows.

Examples of References

Reproduced with permission of the Institute of Food Technologists, Chicago, IL, from Food technology style guide, *Food Technol.* 44(7):129, 1990.

Anonymous

Anonymous. 1985. Vitamin D status of the elderly. Contributions of sunlight exposure and diet. Nutr. Rev. 43:78.

Book

Decareau, R.V. 1985. "Microwaves in the Food Processing Industry." Academic Press, New York.

NRC. 1989. "Recommended Dietary Allowances,: 10th ed. Food and Nutrition Board, Commission on Life Sciences, National Research Council. National Academy Press, Washington, D.C.

Bulletin, Circular

ACSH. 1982. Food additives and hyperactivity. Am. Council on Science and Health, Summit, N.J.

USDA. 1984. Food consumption, prices, and expenditures, 1963-83. 1984 Stat. Bull, No. 713. U.S. Dept. of Agriculture, Washington, D.C.

Chapter in Book

Tilton, E.W. and Burditt, A.K. Jr. 1983. Insect disinfestation of grain and fruit. In "Preservation of Food by Ionizing Radiation," ed. E.S. Josephson and M.S. Peterson, Vol. 3, p. 215. CRC Press, Boca Raton, Fla.

Government Citations

CFR. 1988. Aspartame. Code of Fed. Regs., Title 21, Sect. 172.804. U.S. Govt. Print. Office. Washington, D.C.

FDA. 1988. Food additives permitted for direct addition to food for human consumption; Aspartame. Food and Drug Admin., Fed. Reg. 53:40878.

Journal Article

Mermelstein, N.H. 1987. Sources of information on plastics packaging for food. Food Technol. 41(5):76.

Paper Presented

Solberg, M., Buckalew. J.J., Chen, C. M., and Schaffner, D.W. 1990. Microbial safety assurance for foodservice facilities. Presented at Annual Meeting, Inst, of Food Technologists, Anaheim, Calif., June 16-20.

Patent

Smith, D.P. 1979. Heat treatment of food products. U.S. patent 4,154,861.

Secondary Source

Kritchevsky, D. 1976. Diet and atherosclerosis. Am. J. Pathol. 84:615. Cited in Zapsalis, C. and Anderle Beck, R. (1985), "Food Chemistry and Nutritional Biochemistry," p. 504, John Wiley & Sons, New York.

Lee, T.W. 1986. Quantitative determination of medium chain triglycerides in infant formula by reverse phase HPLC. J.Am. Oil. Chem. Soc. 63:317. Cited in Food Sci. Technol. Abstr. 18(10):115 (1986).

Thesis

Keenan, M.C. 1983. Prediction of thermal inactivation effects in microwave heating. M.S. thesis, Univ. of Massachusetts, Amherst.

Unpublished Data/Letter/Manuscript

Mattern, P.J. and Shoemaker, V.M. 1985. Personal communication. Univ. of Nebraska, Lincoln.

Winters, D.A. and Batt, C.A. 1986. Unpublished data. Cornell Univ., Ithaca, N.Y.

9. *Appendix*: This section should include the following information: (a) sample calculations (not just formulas but substitution into those formulas with proper units), (b) original data sheets, and (c) any other item of secondary interest.

Laboratory 1

UNITS, FUNCTIONS AND DATA PRESENTATION

SUMMARY

This laboratory is designed to present the concept of units, dimensions, functions, and data presentation techniques as utilized by food engineers and processors in dealing with physical quantities of interest. Fundamental and derived units are discussed. A detailed discussion of the SI system is followed by examples. Common mathematical functions and data presentation techniques are graphically illustrated.

1.1. BACKGROUND

In engineering analysis, the part of the universe that is selected for consideration of the changes that may occur within it under varying conditions is called a system, and the region around it is known as the surroundings. The surface dividing the system from its surroundings is termed as the boundary. The formulation, solution, and interpretation of the analysis requires that the physical quantities of the systems are expressed in appropriate units; functional relationship among variables are established; and tables, figures, and graphs are used to make concise presentations of decision-making information.

1.2. UNITS AND DIMENSIONS

1.2.1. Definitions

Dimension: The basic concept used to designate a physical quantity such as time, mass, and length. Both sides of an equation are required to be dimensionally consistent.

Unit: The means of expressing the magnitude of the dimensions (e.g., centimeters or feet for length, hours for time, kilograms for mass).

Base Units: Units that are used to express only one dimension (e.g., the International System (SI) of Units has seven base units--meter (abbreviated m), kilogram (kg), second (s), ampere (A) for current, Kelvin (K) for temperature, mole (mol) for amount of substance, and candela (cd) for luminous intensity).

Secondary Units: Units that are a combination of base units (e.g., newton (N) for force, joule (J) for energy, and coulomb (C) for electric charge).

Accuracy: The degree to which a measurement agrees with a known standard.

Precision: The degree of deviation of a measurement from the mean.

Significant Digits: The smallest value of the measurement unit that can be consistently reproduced.

1.2.2. Common Systems of Units

The major systems of units commonly used are shown in Table 1.1.

The International System (SI) of Units. To unify the usage of units, symbols, and quantities, several international organizations recommended adoption of the "Systemè International d'Unitès," or the SI system of units. Most data still are recorded in older units and thus it is necessary to be familiar with other unit systems. The SI system consists of seven base units, two supplementary units, and a series of derived units.

The seven base and two supplementary SI units are well-defined, dimensionally independent units. Their definitions, along with their symbols are summarized in Table 1.2.

Units for all other quantities are then derived from the above units. In Table 1.3, 19 SI derived units with special names are listed that were derived from the base and supplementary units without numerical factors.

All other SI derived units, like those in Tables 1.4, are similarly derived in a coherent manner from the 28-base, supplementary, and special-name SI units. Some have been given special names.

Expressing Numbers in SI Units. For use with the SI units there is a set of 16 prefixes (Table 1.5) to form multiples and submultiples of these units. Kilograms is the only SI base unit with a prefix. Because double prefixes are not to be used, the prefixes in the case of mass are to be used with gram (symbol g) and not with kilogram (symbol kg).

Rules for Expressing Numbers

1. Prefix to a unit should be chosen so that the numerical value preferably lies between 0.1 and 1000.
2. The prefix symbol and the unit symbols should not be separated by a space.
3. Multiple and hyphenated prefixes should not be used. It is incorrect to write 1 nm as 1 mμm.
4. The exponent to a unit with a prefix means that the multiple of the unit is also raised to the same power. For example, $nm^3 = (nm)^3 = (10^{-9}\ m)^3 = 10^{-27}\ m^3$, not $10^{-9}\ m^3$.

Table 1.1. Common Systems of Units in Use

	Length	Time	Mass	Force	Energy*	Temperature
Absolute Systems						
CGS (cm-gm-sec)	Centimeter	Second	Gram	Dyne*	Erg, joule, or calorie	K,°C
FPS (foot-lb-sec or English absolute)	Foot	Second	Pound	Poundal*	Foot-poundal	°R,°F
SI	Meter	Second	Kilogram	Newton (N)*	Joule	K
MKS (meter-kg-sec)	Meter	Second	Kilogram	Kilogram	Kilogram-calorie	°C
Gravitational Systems						
British engineering	Foot	Second	Slug*	Pound-weight	Btu, (ft)(lb), (ft)(lb_f)	°R,°F
American engineering	Foot	Second	Pound-mass (lb_m)	Pound-force (lb_f)	Btu or hp	°R,°F

*Unit derived from base units.

Table 1.2. The International System of Units--SI

Quantity	Name	Symbol	Definition
Length	Meter	m	1 650 763.73 wavelengths in vacuum of the radiation emitted in the unperturbed transition between the $2p_{10}$ and $5d_5$ levels in the Krypton-86 atom
Time	Second	s	9 192 631 770 periods of the radiation from the unperturbed transition between the hyperfine levels of the ground state of Cesium-133 atom
Mass	Kilogram	kg	The mass of the international kilgram prototype of platinum-iridium alloy, stored at Sevres, France
Current	Ampere	A	The current in two long parallel wires, of negligible cross section and one meter apart in a vacuum, which gives rise to a magnetic force per unit length on each wire of 2×10^{-7} N/m
Temperature	Kelvin	K	The unit in the thermodynamic temperature scale in which absolute zero is K and the triple point of water is 273.16 K
Amount of substance	Mole	mol	That amount of substance containing the same number of elementary entities as there are atoms in 0.012 kg of carbon-12
Luminous	Candela	cd	The luminous intensity, in the perpendicular direction, of a surface of 1/600 000 m^2 of a black body at the temperature of freezing platinum under a pressure of 101 325 Pa

Supplementary Units

Quantity	Unit	Symbol	Quantity	Unit	Symbol
Plane angle	Radian	Rad	Solid angle	Steradian	Sr

Table 1.3. SI Derived Units with Their Names

Quantity	Name	Symbol	Expression in terms of other units
Frequency	Hertz	Hz	s^{-1}
Force	Newton	N	$kg.m/s^2$
Pressure, stress	Pascal	Pa	N/m^2
Energy work, quantity of heat	Joule	J	N.m
Power, radiant flux	Watt	W	J/s
Electric charge, electric quantity	Coulomb	C	A.s
Electric potential, potential difference, electromotive force	Volt	V	W/A
Capacitance	Farad	F	C/V
Electric resistance	Ohm	Ω	V/A
Conductance	Siemens	S	A/V
Magnetic flux	Weber	Wb	V.s
Magnetic flux density	Tesla	T	Wb/m^2
Inductance	Henry	H	Wb/A
Luminous flux	Lumen	lm	cd.sr
Illuminance	Lux	lx	lm/m^2
Celsius temperature*	Degree Celsius	°C	K

*In addition to the thermodynamic temperature (symbol *T*) expressed in Kelvins, use is also made of Celsius temperature (symbol *t*) defined by the equation

$$t = T - T_o$$

where T_o = 273.15 K by definition. One degree Celsius is equal to one degree Kelvin in amount, but to convert "degree Celsius" to Kelvin, 273.15 is added to it. A temperature interval or a Celsius temperature difference can be expressed in degrees Celsius as well as in Kelvins. For example, if t_1 = 20°C and t_2 = 60°C, then T_1 = 20 + 273.15 = 293.15 K and T_2 = 60 + 273.15 = 333.15 K. The difference $t_1 - t_2 = T_1 - T_2$ = 60 - 20 or 333.15 - 293.15 = 40°C or K, but t_1 or t_2 is not equal to T_1 or T_2, respectively.

Table 1.4 Examples of SI Derived Units Expressed in Terms of Base Units or by Means of Special Names

Quantity	SI Unit/Name	Unit symbol
Dynamic viscosity	Pascal second	Pa.s
Heat flux density, irradiance	Watt per square meter	W/m^2
Luminance	Candela per square meter	cd/m^2
Magnetic field strength	Ampere per meter	A/m
Molar entropy, molar heat capacity	Joule per mole Kelvin	J/(mol.K)
Specific heat capacity, specific entropy	Joule per kilogram Kelvin	J/(kg.K)
Surface tension	Newton per meter	N/m
Thermal conductivity	Watt per meter Kelvin	W/(m.K)
Wave number	1 per meter	1/m

Other Rules

1. The symbols do not change in the plural and are never written with a period except at the end of a sentence.

2. A space must be left between the numerals and the first letter of the symbol.

3. The symbols may be shown by using a solidus (/) between the symbols in the numerator and those in the denominator (10 m/s^2) or by the use of symbols with negative exponents (10 $m.s^{-2}$, but not m/s/s).

4. If a numerical value is less than one, a zero should precede the decimal point.

5. To avoid confusion, spaces must be used instead of commas to divide a long row of digits into easily readable blocks of three, starting from the decimal point (e.g., 12 513.126 102).

6. "Coined" names should not be used (e.g., kilo for kilogram).

7. When the names of units are written out, division is indicated by the word "per" and not by the solidus.

Table 1.5 SI Prefixes

Factor	Prefix	Symbol
10^{18}	Exa	E
10^{14}	Peta	P
10^{12}	Tera	T
10^{9}	Giga	G
10^{6}	Mega	M
10^{3}	Kilo	k
10^{2}	Hecto	h
10^{1}	Deka	da
10^{-1}	Deci	d
10^{-2}	Centi	c
10^{-3}	Milli	m
10^{-6}	Micro	μ
10^{-9}	Nano	n
10^{-12}	Pico	p
10^{-15}	Femto	f
10^{-18}	Atto	a

1.2.3. Operations with Significant Digits

In making measurement of a quantity of interest, the figures that are jotted down are called significant digits. Significant figures include all nonzero digits and nonterminal zeroes in a measurement. In whole numbers, terminal zeroes may be significant when specified. In decimal fractions, terminal zeroes are significant whereas zeroes preceding nonzero digits are not. Thus the value 101 contains three significant digits. A measurement of 0.01010 contains four significant figures. If not specified, it would be difficult to tell how many significant digits are contained in the number 256 000. The 6 may well be the measured digit and the three zeroes are needed to place the decimal point. Alternatively, all three zeroes may be significant. To avoid uncertainty, such numbers be expressed in scientific notation in such a way that all significant digits are included in the number appearing the power of 10. Consequently, 2.56×10^5 has three significant figures, 2.560×10^5 has four.

Mathematical operations with measured quantities should result in numbers having the same precision as the least precise of the quantities involved. To add or subtract measured quantities, first all values should be rounded off to correspond in precision to the least precise value involved. For example, 10.786 m + 2.2 m + 6.13 m = 10.8 m + 2.2 m + 6.1 m = 19.1 m.

In multiplication or division, retain in the product or quotient only as many significant figures as are contained in the least precise number. Suppose it is desired to multiply 2.22 m by 3.1 m to find the area of a given rectangle. The answer will be 6.9 m^2.

Rounding of Data

The procedure is as follows:

1. When the first digit discarded is less than 5, the last digit retained should not be changed (e.g., 2.652 rounded to 2.65).

2. When the first digit discarded is greater than 5, or if it is a 5 followed by at least one digit other than zero, the last figure retained should be increased by one unit (e.g., 3.158 502 rounded to 4 digits is 3.159).

3. When the first digit discarded is exactly five, followed only by zeros, the last digit retained should be increased by one if it is odd, but no adjustment made if it is even.

1.2.4. Dimensional Analysis and Conversion of Units

The dimension of a physical quantity is a term that describes the nature of the quantity under consideration. The dimensions of such physical quantities as mass, length, time, temperature, velocity, force, energy, etc. are different. The dimensions of derived quantities can be expressed in terms of base and supplementary units. It is very necessary that the magnitude of a quantity must always be accompanied with its appropriate unit. The interconversion of units is performed by writing a dimensional equation for the conversion. In additions and subtractions, only like terms can be added or subtracted. For example, $(3\ ft)^2 = 9\ ft^2$; $(10\ kg)(2\ m)(3/s^2) = 60\ kg.m/s^2 = 60\ N$; 15 kg - 3 kg = 12 kg; but 15 kg - 300 g cannot be solved unless their units are made alike. Dimensional consistency of an equation does not guarantee that the equation is correct, but inconsistency is a sure sign of incorrectness in the equation.

Conversion of physical quantities from one system of units to another is possible if either the conversion factors for the fundamental units are known or the direct conversion factors of units involved are defined. Comprehensive tables of conversion factors for various units are widely available. Some partic-ularly useful conversion factors are shown in Appendix A, which should suffice for most of the needs of the laboratory work.

Conversion of Mass and Force. Mass is the quantity of matter in a system under consideration. In SI units, mass is expressed in kilograms. In the American engineering or FPS system, the unit for mass is the pound mass (designated lb_m). Two other units for mass are the pound mole (lb_m.mole) and gram mole (g.mole). These units express the quantity of a material whose mass in pounds mass and grams mass, respectively, is equal to the molecular weight of the substance.

Force is "what is required to displace an object." Remember that weight always refers to a force. The SI unit of force is newton (N). The unit of force in the American engineering system is pound force (designated lb_f), which is the weight of one pound mass on the surface of the earth, at sea level.

The relationship between force and mass is given by the Newton's second law of motion: $F = m\,a$, where F = force, m = mass, and a = acceleration. Expressing the above relationship in the form of a dimensional equation in FPS units

$$lb_f \; [=] \; lb_m \cdot (ft/s^2)$$

where the symbol [=] means "has the units of."

The two sides of the foregoing equation are made dimensionally correct using a dimensional constant. Since the standard value for acceleration due to gravity is 32.1740 ft/s^2 (9.81 m/s^2), the dimensional constant would have a value of 32.1740 and would be a denominator with dimensions of $(ft/s^2)(lb_m/lb_f)$.

The dimensional constant is g_c and relates force and mass as follows:

$$F = m.a/g_c, \quad g_c = 32.1740 \; lb_m.ft/(lb_f.s^2)$$

As an example, a mass of 10 lb_m will exert a force of about 9 lb_f at a location where the acceleration due to gravity is 29 ft/s^2.

In the SI system, the unit of force, newton (N) is defined as the force that will accelerate a one kilogram mass at the rate of 1 m/s^2. Using Newton's second law:

$$F = m.a \quad \text{or} \quad 1 \text{ N} = 1 \text{ kg} \; . \; 1 \text{ m/s}^2$$

Now consider the force that must act on a one kilogram mass near the earth surface. The acceleration due to gravity is 9.81 m/s^2.

$$F = (1 \text{ kg})(9.81 \text{ m/s}^2) = 9.81 \text{ kg.m/s}^2 = 9.81 \text{ N}$$

Thus, the earth's gravitational field must exert a force of 9.81 newtons on one kilogram mass. Therefore, the kilogram mass weighs 9.81 N.

Prelab Questions

Q1. What system of units is commonly used by industry and technologists in the U.S.?

Q2. What are the SI units for the following quantities? (a) force, (b) energy, (c) temperature, and (d) solid angle.

Q3. Express the following numbers in correct SI notations: (a) 0.00001 m, (b) 10,000 N/m^2, and (c) 1,000,000 J.

Q4. Determine the number of significant digits in each of the following measurements: (a) 786 ft, (b) 78.6 m, (c) 0.0786 g, (d) 786.00 s, (e) 78.6 x 10^3 m^3, (f) 0.7860 kg, (g) 786.0 N, and (h) 10^5 Pa.

Q5. Add 18.1 m, 10.05 m, 6.58 m. Subtract 48.95 m^2 from 88.1 m^2.

Q6. The dimensions of a crate are 10.30 ft x 3.5 ft x 4.2 ft. What is its volume?

Q7. Convert 15.5 J to ft-lb_f.

Q8. A deceleration of 25 g (acceleration due to gravity) has been found to be fatal to humans in automobile accidents. Compute the force acting on a man weighing 70 kg at this deceleration.

1.3. MATHEMATICAL FUNCTIONS

When two variables are related in such a manner that the value of the first variable is fixed when the value of the second variable is specified, then the first variable said to be a function of the second. The volume of a meatball is a function of its radius, and the volume of a can is a function of its radius and height. The second variable or set of variables, to which values may be independently assigned is called the independent variable, or argument; and the first variable, whose value is fixed once the value of the independent variable is given, is called the dependent variable, or function.

The symbols $f(x)$, $F(x)$, $g(x)$, etc. are used to denote a function of x, and is read "f of x", not f multiplied by x. Thus, if

$$f(x) = x^2 - 2x + 5$$

then

$$f(0) = 5$$
$$f(1) = 4$$
$$f(2) = 5$$

1.3.1. Functions and Equations

The relationship among variables is expressed by equations. The equations indicate the degree of functional relationship and are defined by the highest power of the variable.

Linear or First Degree Equations. Equations of the form

$$f(x) = c_1x + c_2; \quad \text{or} \quad f(x,y) = c_1x + c_2y + c_3$$

where $c_1, c_2, c_3, \ldots,$ are constants, or

$$Z = 2x - 3; \quad \text{or} \quad Z = 3x + 2y - 6$$

are called first degree or linear equations since all variables are raised only to the first power.

Quadratic or Second-Degree Equations. Consider the equations of the form

$$f(x) = c_1x^2 + c_2x + c_3$$

where c_1 is not zero.

$$f(x,y) = c_1x^2 + c_2y^2 + C_3xy + c_4x + c_5y + c_6$$

where c_1 or $c_2 \neq 0$, or

$$Z = x^2; \quad \text{or} \quad Z = 2x^2 + 5y^2 + 3x$$

where the highest power of the variable x or y is 2. Such equations are known as the quadratic or second-degree equation.

Higher-Order Equations. When the relation between two variables (x and y) cannot be expressed by either the linear (first-order) or quadratic (second-order) equations, it may be possible to express the relation by still higher-order equations such as third-order (cubic) or fourth-order or higher-order expressions.

Power Series. Two variables can be related in a general manner by a power series, such as:

$$Z = c_1 + c_2x + c_3x^2 + c_4x^3 + \cdots$$

It is evident that other expressions discussed earlier are all special cases of the basic power series. For example, if all constant except c_1 and c_2 are zero, the power series becomes a linear expression of the form:

$$y = c_1 + c_2x$$

If all constants except c_1, c_2, and c_3 are zero we get a quadratic equation:

$$y = c_1 + c_2x + c_3x^2$$

A power series can also be written as a sum of terms of the form:

$$a_0 + a_1x + a_2x^2 + a_3x^3 + a_4x^4 + \cdots$$

or $\sum_{n=0}^{\infty} a_n x^n$

where $a_o, a, \ldots$ or a_n are constants

Similarly, equations can relate more than two variables in a number of different manners.

Exponential and Logarithmic Equations. Equations of the general form

$$f(x) = a^x$$

where $a > 0$.

or

$$y = 10^x$$

or

$$y = e^x$$

are called exponential equations. The logarithmic functions are related to exponential function in the following way:

$$\log_a f(x) = x$$

or

$$\log_{10}(y) = x$$

or

$$\log_e(y) = x$$

Logarithmic functions are very important in the study of many food processes.

1.4. PRESENTATION OF DATA

For maximizing the decision-making information contained in measurements made on a system, a medium of communication is required. It is needed to report findings, visualize relationships, and assist in further analysis. Engineers and scientists must use tables, figures, and graphs to concisely present the data so that useful information becomes readily apparent. We will briefly review some of the most common data presentation tools.

1.4.1. Tables

Tables are the most common media of communication in the presentation of data. They are simple and consist of an interrelation between rows and columns. The table can be made more complex by more than one variable in both the rows and columns. Column headings are used as references for what is contained between and within the rows.

1.4.2. Figures

A pictorial presentation of data by means of figures highlights the information more vividly. Commonly used figures include the bar diagram and pie chart (sectogram). The advantage in using figures is their simplicity and visual effect.

1.4.3. Graphs

Graphs are data presentation media combining the row and column features of the table and the pictorial presentation of the figure. The values of the independent variable are plotted horizontally (x axis or abscissa) and the values of the dependent variables are plotted vertically (y axis or ordinate). Let us examine the salient features of good graphs.

The Elements of Good Graphs

A. Title: Identification of the primary relationship

B. Legend:
   - (i) Source: author, date
   - (ii) Conditions: product, equipment, constants, etc.

C. Scales:
   - (i) For nontechnical audiences use linear scales whenever possible
   - (ii) For technical audiences choose scales that permit comparison with fundamental relationships, linearize and/or compress data, and permit easy analysis.
   - (iii) For reports (written and visual) use all the center portion of the graph paper, whenever possible. For 8.5 in. x 11 in. sheets, this is about 5 in. x 8 in. including labels.
   - (iv) Identify axes with quantity and SI units. Use SI units and their prefixes.

D. Data:
   - (i) Use clearly differentiated symbols to represent each data point (e.g., *,□,O).
   - (ii) Consider solid and open forms of symbols to distinguish subsets of data (e.g., ∘, •, □, ■).
   - (iii) Use a key to identify data sets (e.g., 10°C, 20°C, 30°C) within a graph.
   - (iv) Use drawing aids such as the template or computer plotting packages.

E. Lines:
   - (i) Choose the number of lines carefully with the audience and communication vehicle in mind (e.g., 2 or 3 for a visual, <6 for a technical report).
   - (ii) Label each line and distinguish lines with the use of ------, --, --, -·-·-·, ····, and similar forms.
   - (iii) Consider using statistical analysis to avoid introducing personal bias in drawing the lines and to produce smoothing of the data.
   - (iv) Use drafting aids such as straight edges, french curves, flexible curves, or a computer plotting package.

F. Appearance:

   Put emphasis on clarity, completeness, neatness and on good form. These elements are essential in establishing and maintaining good communications with your audience.

Many of the aforementioned features are available in computer software used for graphics. Students are encouraged to use them.

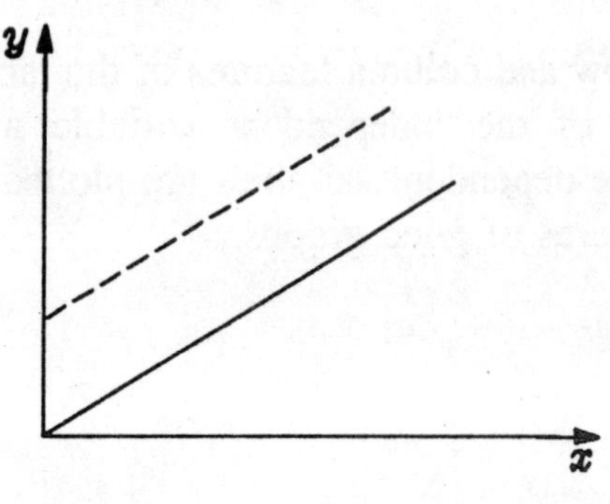

Fig. 1.1 Linear graph.

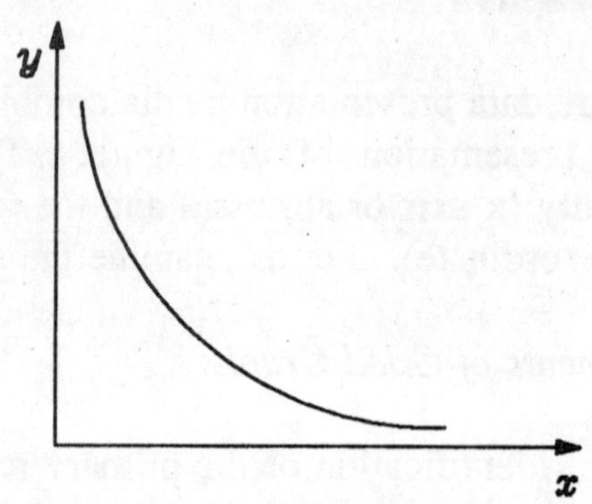

Fig. 1.2 Hyperbolic graph.

The Major Features. The shape of the plotted curve often indicates the mathematical relationship that exists. When the graph is a straight line, the dependent variable *y*, varies directly as the independent variable *x* (Fig. 1.1). A hyperbola indicates that the dependent variable *y* varies inversely as the independent variable $1/x$ (Fig. 1.2). A parabola indicates that the dependent variable *y* varies as the square of the independent variable *x* (Fig. 1.3).

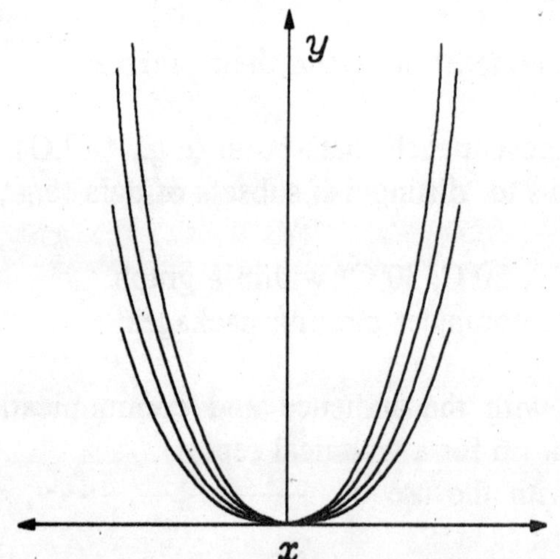

Fig. 1.3 Parabolic graph.

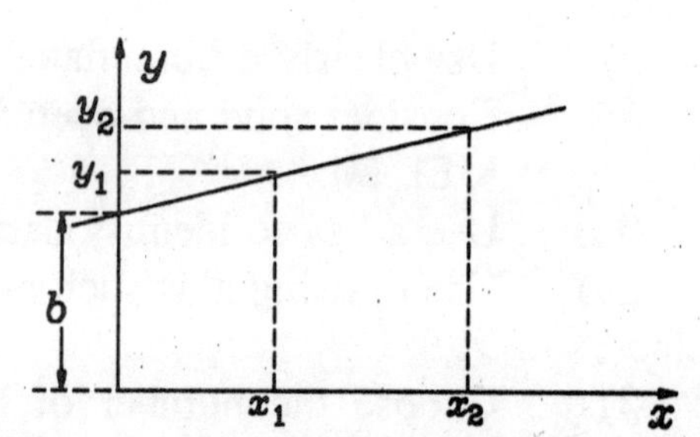

Fig. 1.4 Linear relationship showing slope and intercept.

1. The *equation* for linearized graphs is y = slope (x) + intercept or

$$y = mx + b$$

or

$$y = (\Delta y/\Delta x)\, x + y_o$$

2. The *slope* is the rate of change of one variable with respect to the other. The slope (m), of a plot of y vs. x is $\Delta y/\Delta x$ or $(y_2 - y_1)/(x_2 - x_1)$ where Δ represents the change in the variable for a small increment such as from point (1) to point (2). See Fig. 1.4.

3. The *intercept* is the value at which the dependent variable, y, of a smoothed line intersects the ordinate (line with $x = 0$) (i.e., $y(0)$ or y_o, or b).

4. The *asymptote* is the line where the relationship approaches at limiting values such as x approaches zero or x approaches infinity (Fig. 1.5).

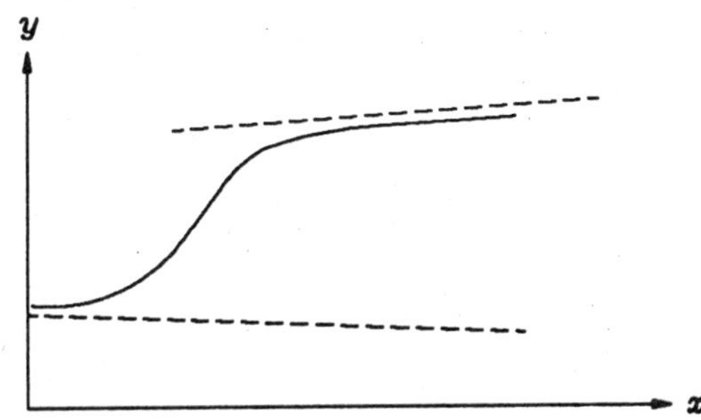

Fig. 1.5 Nonlinear relationship showing asymptotes.

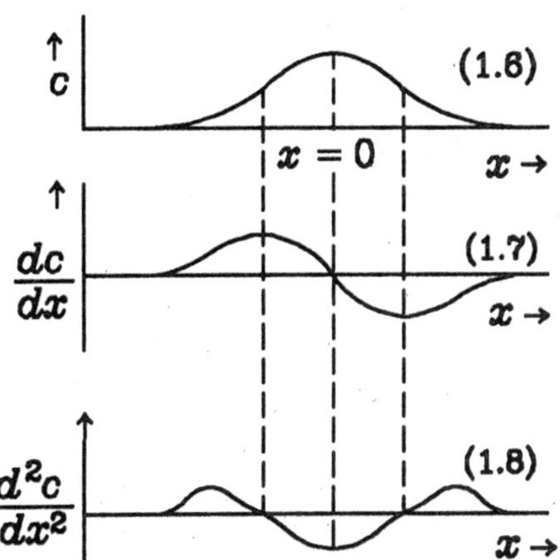

Fig. 1.6, 1.7, and *1.8* show $c(x)$, dc/dx vs. x, and d^2c/dx^2 vs. x, respectively.

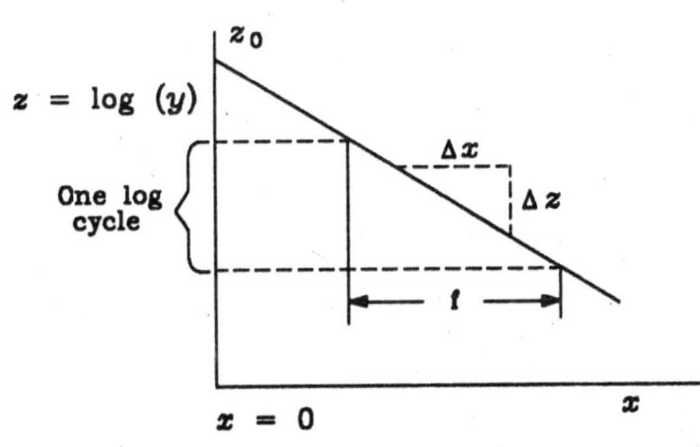

Fig. 1.9 Semilog graph.

Such asymptotic values are found when a potential difference or driving force approaches equilibrium, the state at which there are energy exchanges but no *net* exchange of energy. We encounter *potential* differences such as: temperature, T; pressure, P; concentration, C or chemical potential; voltage, V; kinetic energy (K.E.) = $m.V^2/(2g_c)$.

These changes (i.e., rate processes) occur as a system attempts to reach the same energy level (i.e., equilibrium) after a change in the environment.

Such asymptotic values also are found as systems approach "ideal" conditions such as "ideal" gases, liquids, or solids in which molecular and particle interactions are minimal.

5. Local largest (*maxima*) and smallest (*minima*) values are shown on Fig. 1.6, along with the location (inflection point) at which the slope (first derivative) changes from increasing to decreasing values.

Figure 1.7 shows that at the inflection point the slope is at a maximum (or minimum) and the slopes at maxima and minima are zero.

Figure 1.8 shows that the change in slope (i.e., the second derivative) is zero at the inflection point, negative at maxima, and positive at minima. Inflection points often occur at the value of the abscissa at which the physical mechanism changes (as on a plot of log viscosity vs. $1/T$) or at close to one standard deviation for measurements with normal distribution as for the output "concentration" on a gas chromatograph.

Coordinate Systems

1. A *linear* graph (i.e., a plot of y vs. x) is the most frequent coordinate system as it readily communicates slopes, intercepts, maxima, minima, and inflection points. This system also permits the determination of higher order derivatives from plots (Fig. 1.8) and of lower order derivatives (Fig. 1.7).

2. A *semilog* graph is a plot of log y vs. x. This system is used to condense the y axis or to plot potential y vs. x as in transient heating or product drying (Fig. 1.9)

Here

$$z = \log y = mx + b$$

$$y = 10^{(mx+b)} = \text{antilog}\,(mx + b)$$

$$m = \Delta z/\Delta x \text{ (Fig. 1.9)}$$

$$= (z_2 - z_1)/(x_2 - x_1) = (\log y_2 - \log y_1)/(x_2 - x_1)$$

$$\text{intercept} = z_o = z(x{=}0) = \log y_o = \log y(x{=}0)$$

f = change in x to produce one log cycle change in y

3. A *log-log* graph is a plot of log y vs. log x. This system is used to compress both coordinates as in the analysis of active vs. molar concentration of chemical species or the analysis of transport coefficients for heat, mass, or momentum transfer.

The graph can be represented by the following equation:

$$\log (y) = m \log (x) + b$$

or

$$y = (\text{antilog}\, b)\, x^m$$

that is, $z = m.w + b$

For the new variables $z = \log y$ and $w = \log x$.

For region I (Fig. 1.10):

$$m = \Delta z/\Delta w = \Delta \log y/\Delta \log x$$
$$= (\log y_2 - \log y_1)/(\log x_2 - \log x_1)$$
$$b = \log y \text{ at } \log(x) = 0 \text{ or at } x = 1.0$$

Nusselt number, Sherwood number, friction factor or sedimentation coefficient vs. Reynolds number, or Fourier number vs. Biot number on log-log scale provide linear relationships.

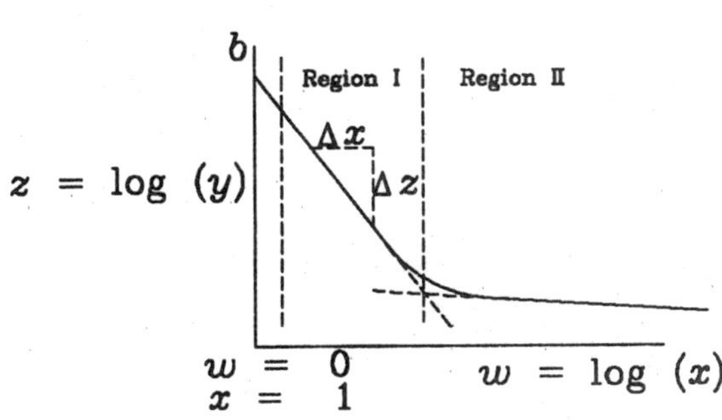

Fig. 1.10 Log–log graph.

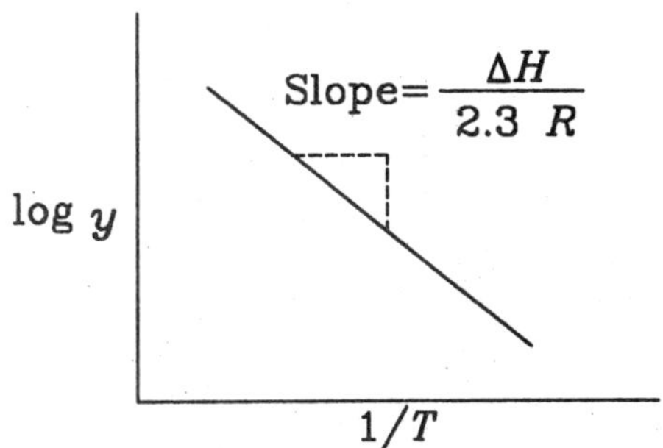

Fig. 1.11 Arrhenius plot of log y vs. $1/T$.

Semilog and log-log graphs are generally made either by taking logarithms of numbers involved and then plotting them on linear graphs using arithmetic scale or using the semilog or log-log graph paper directly.

4. *Arrhenius plots* are graphs of log y vs. $1/T$ and arise from fundamental thermodynamic considerations of rate phenomena such as vapor pressure, viscosity, mass diffusivity, or chemical reaction rates (Fig. 1.11).

These plots are represented by the equation:

$$y = A \exp(-\Delta H/R.T)$$

or

$$\log y = \log A - \Delta H/(2.3R)\,(1/T)$$

where ΔH is the thermal energy required to accomplish the transfer processes, R is the universal gas constant, and A is the frequency factor. Hence, slope = $-\Delta H/(2.3R)$ and intercept = log A.

5. *Probability graphs* are used to determine the arithmetic average (mean) and randomness (standard deviation) of a variable which is normally distributed about a central value (Figs. 1.12 to 1.14).

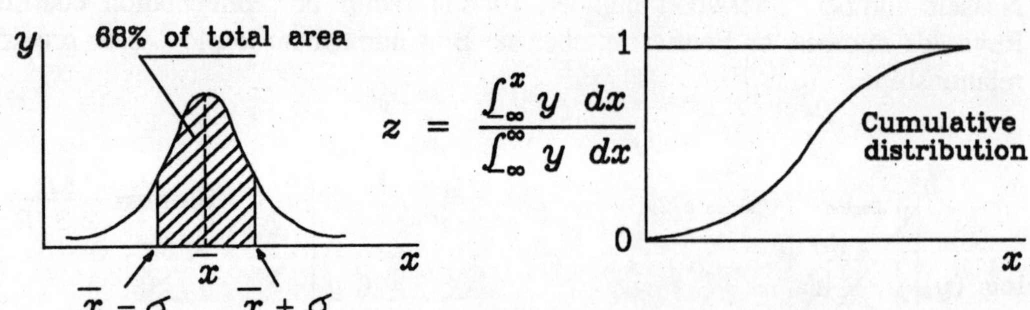

Fig. 1.12 Normal distribution data.

Fig. 1.13 Cumulative distribution of Fig. 1.12.

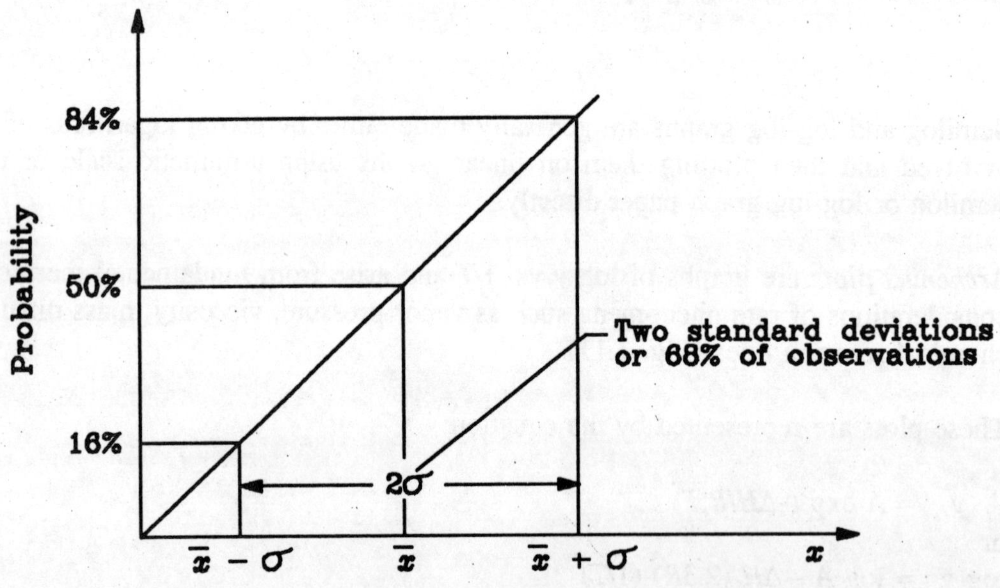

Fig. 1.14 Plot on a probability graph paper. $\bar{x}$: sample mean, σ: standard deviation.

PROBLEM SET

I. *GRAPHICS*

1.1. The active chloride concentration (a_{Cl}) of a formulated food product was measured by a chloride electrode at several levels of chloride concentration (C_{Cl}). The data generated are as follows:

a_{cl}	log a_{Cl}	C_{Cl}	log C_{Cl} (mol/L)
0.0759	-1.12	0.1000	-1.0
0.0081	-2.09	0.0100	-2.0
0.0010	-3.0	0.0010	-3.0
0.0001	-4.0	0.0001	-4.0

a. Plot log a_{Cl} vs. log C_{Cl} on arithmetic coordinates or on log-log graph paper. Then determine the slope and intercept.

b. Determine the exponent m and b for a_{Cl} = (antilog b)$C_{Cl}{}^{m}$

c. The activity coefficient (γ), a property of the system, is defined as $\gamma = (\Delta a_{Cl}/\Delta C_{Cl})$ and is a function of concentration. Determine γ for each of the three intervals and plot as γ vs. log C_{Cl} on arithmetic coordinates. Does the activity coefficient γ approach 1.0 for dilute (i.e., ideal) solution in this product?

1.2. A sample of beans was dried in a thin layer and weighed periodically. The sample was dried to equilibrium, then dried in a vacuum oven to determine the dry solids and mass of water remaining at equilibrium. The data were converted from mass to percent moisture on the dry basis (i.e., (gram of water x 100)/gram of solids). The data are as follows:

Time, min	Moisture Content, %, dry basis
0	10.0
2	9.5
4	8.5
6	7.8
8	6.7
10	5.9
15	4.3
20	2.8
30	2.2
40	1.4
50	0.9
60	0.5
70	0.2

Determine the intercept, the slope and the f-value for moisture diffusion. The apparent moisture diffusivity is $(2.303\, r^2B^2/\pi^2 f)$ where r, the bean radius, is 0.3 cm and B, a shape factor, is 1.0. Calculate the apparent moisture diffusivity.

1.3. A sample of spray dried powder was sieved to determine the size distribution. The mass fraction retained on each sieve was as follows:

Mesh No.	Sieve Diameter in.	Sieve Diameter mm	Mass Fraction Retained
10	--	--	--
14	0.0555	1.4097	0.19
20	0.0394	1.0007	0.32
28	0.0280	0.7112	0.20
35	0.0198	0.5029	0.13
48	0.0140	0.3556	0.07
65	0.0099	0.2514	0.04
100	0.0070	0.1778	0.03

(a) Plot mass fraction vs. log sieve diameter.
(b) Plot cumulative fraction larger than each sieve diameter to log sieve diameter.
(c) Determine the mean diameter and the standard deviation for the high and low sides of the size distribution.

1.4. The following data are expected to follow a linear relation of the form $y = a.x + b$. Plot the data in an appropriate manner and find the straight line relationship.

x	0.9	2.3	3.3	4.5	5.7	6.7
y	1.1	1.6	2.6	3.2	4.0	5.0

1.5. The following data points are expected to follow a functional variation of $y = ax^b$. Obtain the values of a and b from a graphical analysis.

x	1.21	1.35	2.40	2.75	4.50	5.10	7.10	8.10
y	1.20	1.82	5.00	8.80	19.50	32.50	55.00	80.00

1.6. The following data points are expected to follow a functional variation of $y = ae^{bx}$. Obtain the values of a and b by graphical analysis.

x	0	0.43	1.25	1.40	2.60	2.90	4.30
y	9.40	7.10	5.35	4.20	2.60	1.95	1.15

1.7. Compute the volume (L) required to contain 1200 kg of marmalade having a specific gravity of 1.17.

1.8. A mercury manometer attached to a piece of equipment reads 10 cm Hg vacuum. The atmospheric pressure is 76.2 cm Hg. What is the equipment absolute pressure in Pa? Density of Hg = 13546 kg/m^3, $\Delta p = \rho.g.\Delta h$, where Δp is pressure change in Pa, ρ is density in kg/m^3, g = 9.81 m/s^2, and Δh is change in liquid column height.

1.9. Air at 22°C, 85 kPa (abs.) moves with a velocity of 5 m/s through a 0.25-m diameter circular duct. Determine the volume flow rate in m^3/s and the mass flow rate in kg/s. R = 0.287 kJ/(kg.K), use $p.v = R.T$.

1.10. At 90% efficiency, at what constant velocity, in m/s, can a 1-kW motor lift a 100-kg load?

1.11. How many kJ of energy must be removed to cool a 100-L tank of milk from 25° to 4°C? The specific heat of milk is about 3.89 kJ/(kg.K) above freezing and specific gravity is 1.03.

1.12. How many W.h of energy must theoretically be supplied to raise a 50-kg can of milk a distance of 10 m?

1.13. A water manometer is being used to measure air pressure in an air duct. If the manometer shows a 25 cm differential, what is the pressure in kPa? Density of water = 1000 kg/m^3.

1.14. What is the heat generation rate, in kJ/h, of a 1000-W electric heater?

1.15. The meter constant of an electric meter is 5 W.h/disc revolution. What is the rate of power consumption if the disc rotates 18 times in 3 min?

1.16. At the current cost of utility electricity, about 4.0 cents/(kW.h), what is the value of the energy stored in a 12-V, 72 A.h car battery? Remember that power (W) = (V)(A) in D.C. circuits.

1.17. What is the pressure, in kPa, at the bottom of an unpressurized tank, 10 m deep, which holds tomato puree with a specific gravity of 1.06?

1.18. What is the average velocity of a fluid flowing in a 3-cm inside diameter pipe, when the flow rate is 100 L/min? Give velocity in m/s.

Prelab Questions for Laboratory 1

NAME:______________________________ DATE:________________

ANSWER PRELAB QUESTIONS ON THIS SHEET

Laboratory 2

MASS AND ENERGY BALANCES

SUMMARY

This lab provides the application of steady-state conservation of mass and energy equations to analyze food processing operations. It also gives the guidelines to use steam tables.

2.1. BACKGROUND

2.1.1. Conservation of Mass

The conservation of mass equation for a system can be written as:

Total mass or rate of mass, in = total mass or rate of mass, out + mass accumulated or rate of mass accumulation (2.1)

Under steady-state conditions, there should be no mass accumulation in the system. Equation (2.1) will be

Total mass or rate of mass, in = total mass or rate of mass, out (2.2)

Equation 2.2 can be written for the total mass or for each component of a product. This can also be applied to each unit operation or a part of the process (combination of unit operations) or to a complete process. These equations are useful in analyzing complex food processing systems. After writing these mass balance equations, unknown mass or mass flow rate or concentration of a component in a food product can be computed by solving simultaneous equations. The number of independent simultaneous equations should be equal to the number of unknowns.

2.1.2. Conservation of Energy

The principle of the conservation of energy is explained by the first law of thermodynamics: "In any system the energy associated with matter entering the system plus the net heat added to the system is equal to the sum of the energy associated with matter leaving the system, net work done by the system, and the change in system energy," that is,

Energy in = energy out + change in the system energy (2.3)

In equation (2.3), heat and work are included in the total energy. In food processing, most of the time we are interested in steady- state conditions, at which the change in the system energy will be negligible. Thus, for steady-state conditions, the conservation of energy equation can be written as

$$\text{Energy in} = \text{energy out} \tag{2.4}$$

The energy term may contain thermal, hydraulic, kinetic, or potential energy forms. In thermal processing of foods (heating, cooling, freezing, cooking, sterilization, pasteurization, etc.), the thermal energy is more pronounced, and other energy terms may be neglected under certain conditions. If they are not neglected, then they can be calculated using Bernoulli's equation (Lab 5). Two forms of thermal energy terms are generally encountered: (1) sensible heat due to temperature change and (2) latent heat due to phase change. Sensible heat is calculated by

$$\text{Sensible heat} = \text{(mass or mass flow rate).(specific heat).(temperature change from a reference)} \tag{2.5}$$

and the latent heat is computed by

$$\text{Latent heat} = \text{(mass or mass flow rate).(latent heat per unit mass)} \tag{2.6}$$

2.1.3. Phase Diagram and Property Tables

The properties (specific volume, enthalpy, entropy, etc.) of common substances (water, refrigerants like Freon-12), as a function of temperature and pressure, have been measured; and those for numerous pure substances have been catalogued in tabular form. A complete set of tables is that for water, called the "steam tables."

Water or any other pure substance can exist in three forms or phases: solid, liquid, and gaseous (vapor). A substance can exist in more than one phase at any one time. Corresponding to any pressure is a temperature at which the liquid phase can coexist with the vapor phase. Such pressure and its corresponding temperature are called the "saturation pressure" and "saturation temperature," respectively (e.g., water at 101.325 kPa has a saturation temperature of 100°C; the saturation pressure corresponding to the saturation temperature of 100°C is, therefore, 101.325 kPa). For every pure substance definite relationships exist between temperature and their corresponding saturation pressures, not only for the liquid-vapor phases but also for the solid-liquid and solid-vapor phases as well.

Figure 2.1 shows the phase diagram (*P-T*) for a pure substance. The following is the discussion on this phase diagram:

1. *Critical point*: The liquid-vapor saturation curve ends at the critical point, at which (a) the liquid and vapor phases are intermixed, and (b) there is no vaporization process.

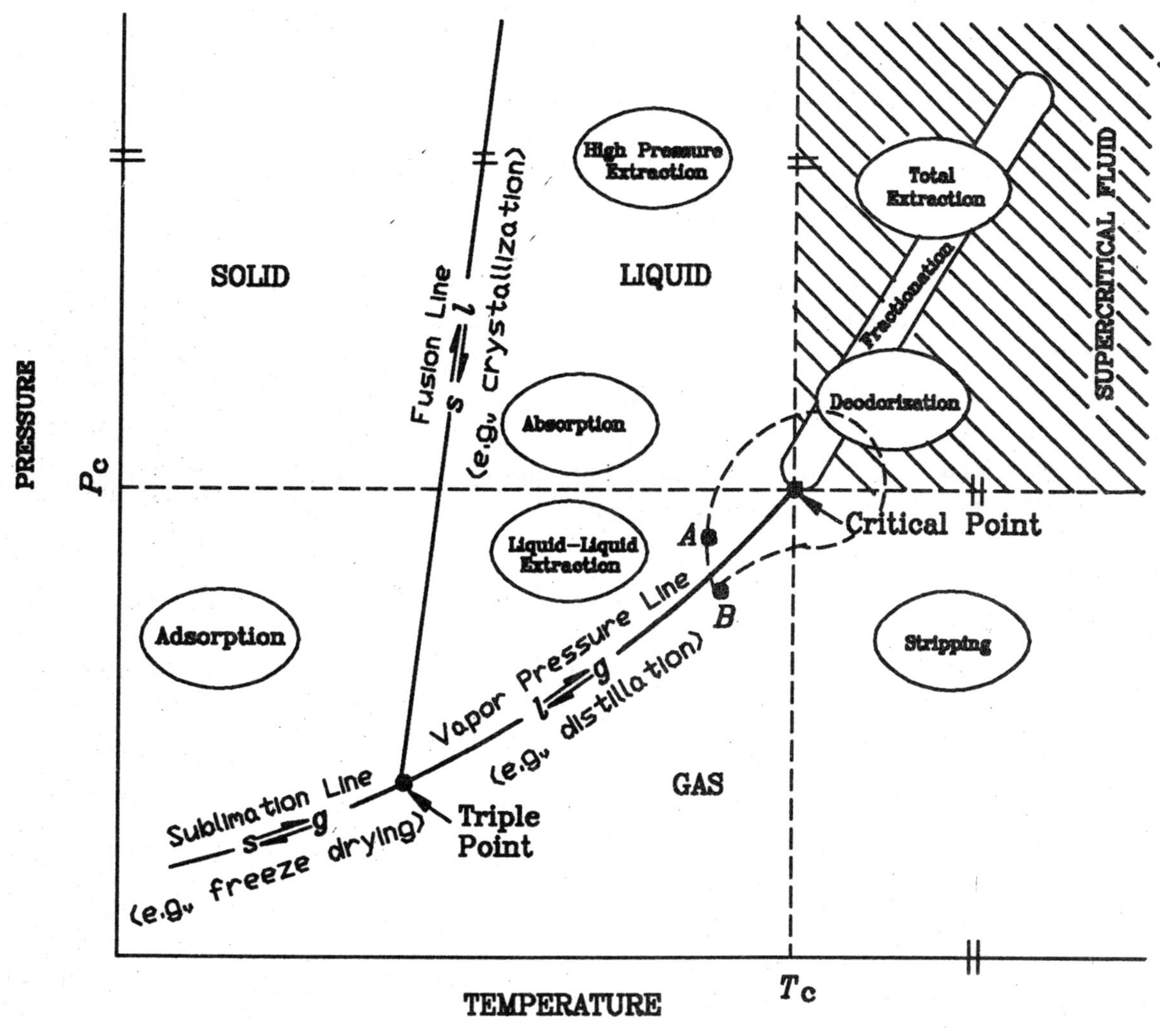

Fig. 2.1 Phase diagram for a pure substance.

2. *Liquid-vapor saturation (vaporization) curve*: Between triple and critical points is liquid-vapor saturation or vaporization curve. A certain amount of heat (latent heat or heat of vaporization) is required to vaporize a unit mass of pure substance in the liquid phase. The heat of vaporization decreases as the critical point is approached. Two phases can coexist at any point on a saturation curve. When only liquid exists at the saturation pressure and temperature, the liquid is called a saturation liquid. When only vapor exists at the saturation pressure and temperature, the vapor is saturated vapor.

3. *Triple point*: The intersection of the three saturation curves (liquid-vapor, solid-vapor, and solid-liquid) is the triple point. All three phases of the pure substance exist in equilibrium at these conditions of pressure and temperature.

4. *Solid-liquid saturation (fusion) curve*: Solid and liquid phases coexist in equilibrium.

5. *Solid-vapor saturation (sublimation) curve*: Solid and vapor phases coexist in equilibrium.

6. *Subcooled or compressed liquid*: In this region, only the liquid phase exists. The temperature in the liquid phase is less than the saturation temperature for a given pressure, hence it is "subcooled liquid." The substance is also called "compressed liquid" because its pressure is greater than the saturation pressure at a given temperature.

7. *Superheated vapor*: In this region, only the vapor phase exists. The temperature is greater than the saturation temperature for the existing pressure.

8. *Quality (x)*: In the two phase regions, since pressure and temperature are not independent and thus another property is required to fix the state of the pure substance (e.g., on the vaporization curve), the part of the substance may be saturated liquid and part saturated vapor. The term "quality" is used for the vaporization curve, indicates what percent or fraction of the mixture is vapor. If $x = 0$, all the substance is saturated liquid, and if $x = 1$, all of the substance is saturated vapor.

The properties of the liquid-vapor mixture are calculated by the following equations:

$$v = v_f + x.v_{fg} \tag{2.7a}$$
$$u = u_f + x.u_{fg} \tag{2.7b}$$
$$h = h_f + x.h_{fg} \tag{2.7c}$$
$$s = s_f + x.s_{fg} \tag{2.7d}$$

where v is specific volume, m^3/kg; u is internal energy, kJ/kg; h is enthalpy, kJ/kg; s is entropy, kJ/(kg.K); subscript f stands for liquid, and subscript fg for the difference in vapor and liquid properties (e.g., $v_{fg} = v_g - v_f$). The h_{fg} is known as the latent heat of vaporization, and h_f is the sensible heat of the liquid phase, whereas h_g is the sum of sensible and latent heats.

2.1.4. Steam Tables

The properties (v, u, h, s) are given in steam tables for various phases of water and their mixture. The first table is generally for saturated water. It has saturated temperature in the first column, and similar tables are available that have saturated pressure in the first column. These provide the properties of water on the liquid-vapor saturation curve. Linear interpolation is used to estimate the properties between two temperatures or pressures. Only one parameter (T or P) is sufficient to determine the properties on saturation curve. Other tables provide the properties of superheated water vapor at different temperatures and pressures. Both temperature and pressure are needed to locate the properties in this region. Similar tables are available for the properties of compressed or subcooled liquid. Again, both temperature and pressure should be specified. These tables are given in Appendix B. The computer program "Table" also provides these properties.

Prelab Questions

Q1. Describe the conservation of mass equation for steady-state conditions.

Q2. Describe the conservation of energy equation for steady-state conditions.

Q3. Differentiate between kinetic and potential energies.

Q4. Differentiate between latent heats of vaporization and fusion.

Q5. Define "saturated water" in liquid and vapor forms.

Q6. Differentiate between "boiling temperature" and "saturation temperature."

Q7. Describe conceptually "compressed liquid."

Q8. Define steam quality.

2.2. PROBLEMS

Solve each of the following step by step. Include units on all answers.

2.1. Conversion of raw potatoes to chips involve several steps. In a chip manufacturing operation it was found that if 10.0 kg potatoes are peeled and eyed, a mass loss of 20% results. After chipping, the potatoes are fried and then found to weigh 7.5 kg. Of the final mass of chips, oil absorbed during frying accounts for 10%. Compute the yield after (a) peeling and eyeing and (b) frying operations.

2.2. Tests show that a raw food coming to a drying plant contains 70% water. After drying, it is found that 65% of the original water has been removed. Compute (a) the water content of the dried food and (b) the mass of water removed per unit mass of wet food.

2.3. During formulation of a new food product, 20 kg of one component (A) containing 40% solids are mixed with an unknown amount of another component (B) containing 70% solids. If 130 kg of the food product are desired, compute the amount of B component necessary. What is the composition, in % solids, of the new product?

2.4. A vegetable after drying to a moisture content of 5% dry basis (weight basis) is to be packaged containing 0.6 kg of dry matter. It is desired to reduce the moisture content from 5% to 2% by placing a desiccant in each package with the vegetable. The desiccant absorbs moisture from the vegetable and when equilibrium is reached, the desiccant contains 10 times the moisture content of the vegetable. If the desiccant initially has zero moisture, what mass of desiccant is required for each package (m.c.d.b. = mass of water/mass of dry matter)?

2.5. Pectin enzymes of the tomato pulp are being deactivated by heating with steam mixing. The initial concentration of total solids in the pulp is 5.7%. The pulp is heated by mixing saturated steam at 105°C, and being heated from 20 to 87°C. Calculate the concentration of total solids. The specific heat of solids is 2.2 kJ/(kg.K) and of water 4.19 kJ/(kg.K).

2.6. A 100-lb (45.36 kg) crate of apples was initially atop a 20-ft (6.1 m) high truck. The crate suddenly fell down and hit the ground. Taking the ground as reference, compute (a) the initial kinetic and potential energies of the crate, (b) the final kinetic and potential energies contained in the crate, and (c) if all the initial potential energy contained in the system is to be converted into heat, how many Btu (or J) would this equal to?

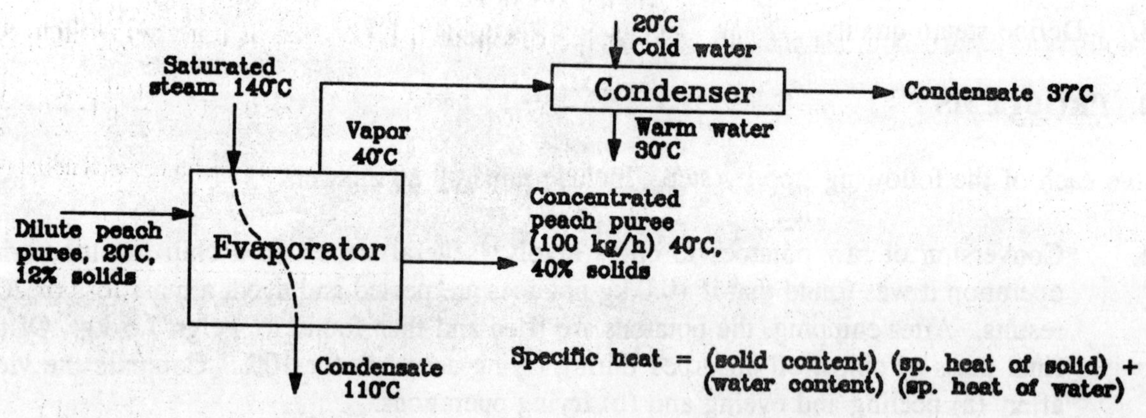

Fig. 2.2 **Schematic diagram for Problem 2.7.**

2.7. A peach puree is being concentrated in a continuous vacuum evaporator at a rate of 100 kg/h. The feed has a temperature of 20°C and a total solids content of 12%. A product of 40.0% total solids is withdrawn at a temperature of 40°C and condensate leaves the condenser at 37°C.

(a) Calculate the flow rates of the product and condensate streams.

(b) If saturated steam (140°C) is condensing at 110°C is used, calculate the steam consumption in kg/h. The specific heat of the solid material is 2.10 kJ/(kg.K), and specific heat of water is 4.19 kJ/(kg.K)

(c) Cooling water enters the condenser at 20°C and leaves at 30°C. Calculate the cooling water flow rate if the consensate leaves at 37°C. Use steam tables for the enthalpies of steam, water and vapor.

2.8. A test with a microwave oven showed that when 1 kg of water was heated in the oven for 200 s the temperature of the water increased from 25°C to 7°C and 10 g of water had evaporated. During standby, the oven consumed 0.2 kW and during heating the consumption rose to 2.5 kW. The megatron was found to have a conversion efficiency of 70%. What is the thermal efficiency of the heating process?

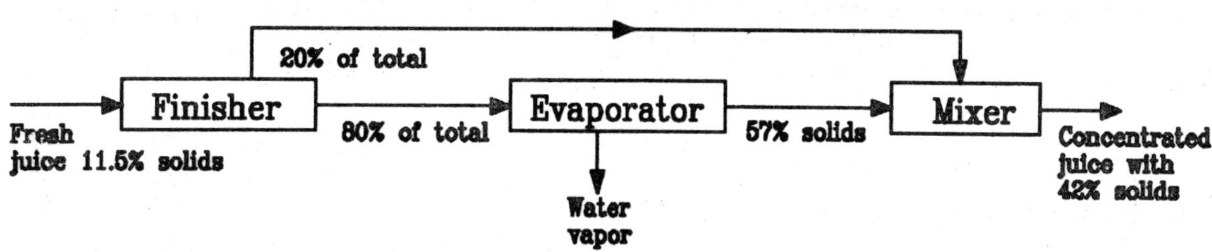

Fig. 2.3 Schematic for Problem 2.9.

2.9. Orange juice extract containing 11.5% solids is being separated into pulpy juice (20% by weight) and strained juice using a finisher. The strained juice in concentrated to 57% total solids by passing through a vacuum evaporator. The pulpy juice is mixed with the concentrated juice to achieve the desired solids concentration of 42% (Fig. 2.3). Calculate (i) the concentration of solids in (a) pulpy juice and (b) strained juice and (ii) the mass of water evaporated per kg of extracted juice.

2.10. Determine the specific volume of water at (i) $T = 250°C$, $P =$ 1.0 MPa, (ii) $T = 200°C$, $X = 0.70$, and (iii) $T = 200°C$, $P =$ 1.0 MPa.

2.11. Consider steam at 300°C, 0.4 MPa flowing through a 0.050-m-diameter pipe with an average velocity of 10 m/s. Determine the flow rate of steam in kg/s.

2.12. We want to prepare condensed milk by boiling off part of the water. Neglecting the slight increase in boiling temperature of milk over pure water, estimate the gauge pressure that must be maintained in a vacuum pan evaporator for the evaporation temperature to be (a) 100°C, (b) 80°C, (c) 60°C, and (d) 40°C. The barometer reading is 76 cm Hg (1 cm Hg = 1333 Pa).

2.13. The pasteurizer plates are heated by saturated steam and the steam temperature is 5°C greater than the maximum product temperature, determine the steam gauge pressure in kPa, for maximum product temperatures of (a) 120°C and (b) 140°C. Assume that the barometer reading is 76 cm Hg.

Prelab Questions for Laboratory 2

NAME:______________________________ **DATE:**______________________

ANSWER PRELAB QUESTIONS ON THIS SHEET

LABORATORY 3

RHEOLOGICAL PROPERTIES OF FLUID FOODS

SUMMARY

A laboratory experiment for measurement of the rheological properties of both Newtonian and non-Newtonian fluids is described. The fundamental equations for determining the rheological parameters from measurements using both capillary and rotational viscometers are provided.

3.1. BACKGROUND

Rheology is broadly defined as the study of the deformation and flow of materials. Characterization of the rheological behavior of materials from their shear stress and shear strain relationship (often called the flow curve) and quantitative understanding of their behavior are of particular interest in material handling and process design. The rheological properties determine such processing parameters as the shape of the velocity profile, the magnitude of the pressure drop and the piping design, and the pumping requirements for a fluid transport system. These properties are also utilized to evaluate the power requirement for agitation, mixing, and blending operations and to determine the amount of heat generation (viscous heating) during extrusion cooking of foods. In product development, rheological properties are utilized to impart desirable characteristics to the end product. Rheological characteristics also influence heat transfer, transfer of mass, mixing, and sedimentation in fluid foods. In some instances, the fluid consistency may be altered by high shear rates encountered in these processes. The fluid resistance can also be used along with diffusion, sedimentation, and electrophoretic mobility to characterize the size and shape of macromolecules in model food systems.

3.1.1. Types of Rheological Behavior

Newtonian Behavior. Food materials range from simple fluids like dilute aqueous solutions, milk, and vegetable oils to complex systems such as meat batters and tomato paste. Fluids that exhibit a linear relationship between shear stress and shear rate passing through the origin are called Newtonian. These fluids are said to follow Newton's law of viscosity, defined by the following equation in the Cartesian coordinate convention.

$$T_{yx} = - \mu \frac{dv_x}{dy} \tag{3.1}$$

where T_{yx} is the shear stress exerted in *y*-direction as a result of laminar flow velocity in *x*-direction; dv_x /dy is the shear rate (γ) or normal velocity gradient (gradient of *x*-component of velocity in *y*-direction); and μ is the viscosity, Newtonian viscosity, or dynamic viscosity.

In Couette flow under no-slip conditions, the shear stress can be defined as the shear force divided by the area perpendicular to the force (F/A). The shear strain (γ) is defined as the ratio of the deformation of a differential element in the x-direction to that in the y-direction (dy). The shear rate then becomes

$$\dot{\gamma} = \frac{d\gamma}{dt} = \frac{d}{dt}\left(\frac{dx}{dy}\right) = \frac{d}{dy}\left(\frac{dx}{dt}\right) = \frac{dv_x}{dy}\text{, \textit{the velocity gradient}}$$

The kinematic viscosity is defined as $\upsilon = \mu/\rho$, where ρ is the fluid density.

Non-Newtonian Behavior. A large number of materials, however, do not follow Newton's law of viscosity. Their shear-stress vs. shear-rate plots either are not linear or do not go through the origin or both. Such materials are generically known as non-Newtonian. There are several classes of non-Newtonian materials.

One class of non-Newtonian materials is called the Bingham plastics. These materials require the application of a finite shear stress, τ_o (called the yield stress or yield value), for the flow to initiate and beyond that they behave much like Newtonian materials. Examples of such behavior include certain fruit and vegetable purees, suspensions, and toothpastes. The defining relationship between shear stress and shear rate for these materials is given by the following two-parameter model.

$$T_{yx} = -\mu_b\left(\frac{dv_x}{dy}\right)+\tau_o, \qquad T_{yx} > \tau_o \tag{3.2}$$

where μ_b is the Bingham viscosity and τ_o is the yield stress. The relationship between shear stress and shear rate of the non-Newtonian materials is most commonly described by a two-parameter power law model of the form

$$T_{yx} = -m\left(\frac{dv_x}{dy}\right)^n = -m\left(\frac{dv_x}{dy}\right)^{n-1}\left(\frac{dv_x}{dy}\right) \tag{3.3}$$

where m is the consistency coefficient and n is the flow behavior index (dimensionless). The non-Newtonian materials are further classified on the basis of deviation of n from unity. When $n < 1$, the material is called pseudoplastic or shearthinning and if $n > 1$, the material is referred to as dilatant or shear thickening. Of course, for $n = 1$, the power law equation reduces to the Newton's law of viscosity.

Because the ratio of shear stress and shear rate for non-Newtonian materials is not a constant value, their flow behavior cannot be expressed by a single value as with the Newtonian behavior. To distinguish this, the ratio of shear stress to shear rate for non-Newtonian materials is designated apparent viscosity (μ_a) as follows:

$$\mu_a = T_{yx}/(-\frac{dv_x}{dy}) = -m(\frac{dv_x}{dy})^{n-1} \tag{3.4}$$

It is important to note that, for a reported value of apparent viscosity of a non-Newtonian material to be used, the shear rate or shear stress used in determination should also be given.

Pseudoplastic materials exhibit a decrease in apparent viscosity with increasing shear rate, equation 3.4. To this big group of materials belong many such common food items as tomato paste, banana puree, apple sauce, most food doughs, and many gum solutions. Pseudoplastics contain asymmetric molecules or particles that could orient during flow and quickly attain random orientation at rest due to Brownian scale forces. Many of these systems also involve many low-level interactions such as hydrogen bonding, hydrophobic bonds, dipole interactions, as well as ionic interactions. Not all interactions would be expected to be broken at energy input rates of low shear. These interactions would lead to structures similar to those originally present if the shear energy rates were removed. The apparent viscosity of dilatant materials increases as shear rate increases. Dilatant materials are encountered rarely, but one familiar example is aqueous suspension of starch. Highly concentrated suspensions such as ceramic slurries and silicones also exhibit this behavior. In dilatant materials, shearing is expected to increase the probability of particle-particle interactions and lead to the development of foamlike structures and to an increase in the extent of particle hydration. Physical dilation (volume enlargement) due to physical displacement of particles from a compact (rhombic) orientation to open orientation (cubic) is probably not a mechanism of foods. For many food products, dilatant behavior is found only in the initial mixing, forming, or transport operations with real plastic behavior found thereafter.

Mixed type or power-law with yield stress or Bingham pseudo-plastic fluids exhibit behavior similar to pseudoplastic but require some yield stress to cause flow. These can be defined mathematically by Herschel-Bulkley or Casson's equation. The different types of flow behavior discussed above are summarized graphically in Fig. 3.1.

3.1.2. Effect of Temperature on Flow Behavior

Temperature is one of the key variables affecting the rheological behavior of materials and its effect must be known since most foods are subjected to different temperatures during processing, as in pasteurization, extrusion cooking, and other unit operations. The Arrhenius relationship is generally used to relate temperature to rheological parameters like viscosity, apparent viscosity, and the consistency coefficient.

$$\mu_a = A \exp(-E_a/R.T) \tag{3.5}$$

where A is a constant, E_a is the activation energy of flow, R is the gas constant, and T is the absolute temperature. The A and E_a parameters are determined by measuring μ_a at several temperatures.

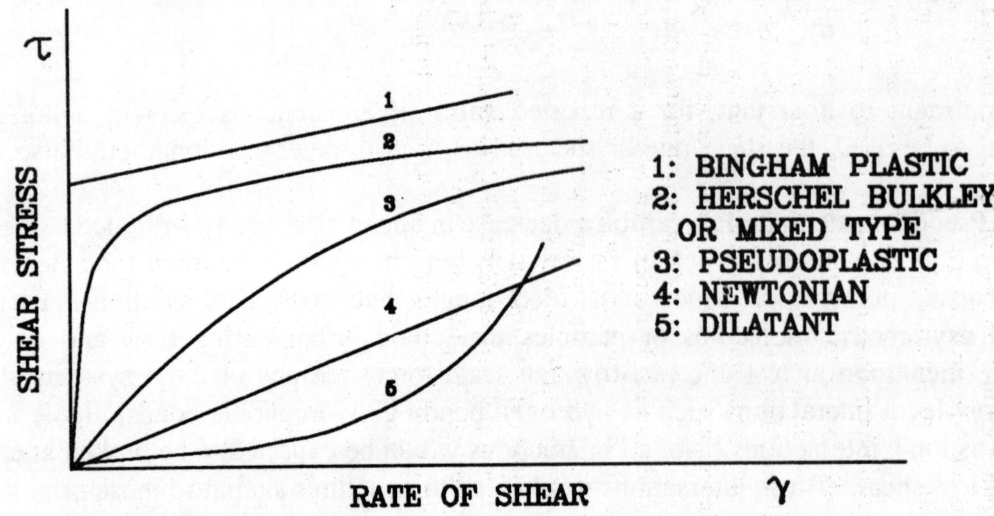

Fig. 3.1 **Shear stress vs. shear rate relationships for various types of fluids.**

A temperature change of just 1° can change viscosity as much as 10% to 15%. It thus essential to control the temperature of the material under test and to report the test temperature.

Prelab Questions

Q1. Define rheology. Why is rheology important? What is shear stress?

Q2. Define viscosity in terms of shear stress and velocity gradient.

Q3. What characteristics of a liquid does viscosity measure?

Q4. What is the effect of temperature on viscosity?

Q5. Define the following terms: (a) Bingham plastic, (b) pseudoplastic, (c) dilatant.

Q6. What is a non-Newtonian liquid? What mathematical relation is generally used to express its behavior?

Q7. What are the units of viscosity in the SI and cgs systems?

3.2. MEASUREMENT OF RHEOLOGICAL PARAMETERS

The Newtonian viscosity of fluids can be calculated from the measured ratio of shear stress and shear rate. One of various types of viscometers such as capillary, falling ball, rotational, and cone-and-plate are generally used. For nonNewtonian materials there are several methods of measurements. To obtain the power law parameters to describe their behavior involves the measurement of shear stress and shear rate relationship using a rotational or a cone-and-plate viscometer.

3.2.1. Capillary Viscometer

The experimental setup used in capillary flow measurements are of different types. One type of capillary viscometer is shown in Fig. 3.2. The ratio of the reservoir bulb radius to that of the capillary should be greater than 10 so that the pressure drop due to the flow in the reservoir can be neglected.

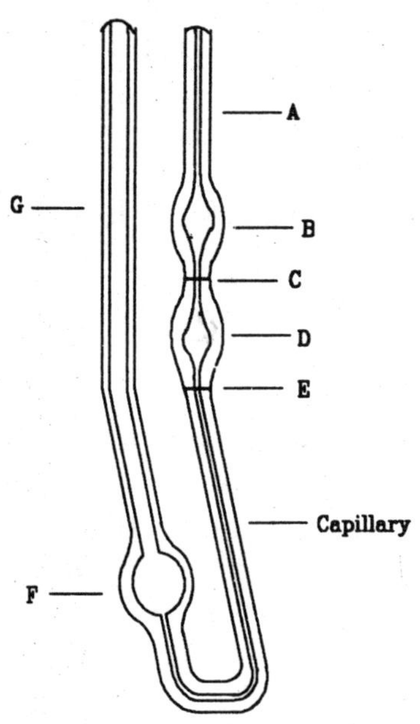

Fig. 3.2 Canon Fenske capillary viscometer. A: arm, B: small bulb, C: etched mark, F: large bulb, G: arm.

For steady laminar flow of Newtonian fluids in a circular capillary, the classical Hagen-Poiseuille equation (3.6), provides the basis for the measurement of viscosity.

$$\dot{V} = \frac{\Delta P.\pi.R^4}{8\mu.L} \tag{3.6}$$

where V is the volumetric flow rate, ΔP is the pressure drop across the capillary, R is the capillary radius, μ is the fluid viscosity, and L is the capillary length.

However, for a mean static fluid height h, the pressure drop is given as

$$\Delta P = \rho g.h \tag{3.7}$$

where ρ is the fluid density and g is the acceleration due to gravity. Substitution of equation 3.7 into equation 3.6 and rearrangement gives

$$\frac{\mu}{\rho} = \nu = \frac{\pi.R^4.g.h}{8.L.\dot{V}} \tag{3.8}$$

where υ is the kinematic viscosity of the fluid. If volume V of the fluid flows through the capillary in time t, then

$$\dot{V} = \frac{V}{t} \tag{3.9}$$

Now equation 3.8 can be rewritten as

$$\nu = \left(\frac{\pi R^4 g.h}{8L.V}\right)t \tag{3.10}$$

When the viscometer shown in Fig. 3.2 is allowed each time to drain between etched marks C and E, the volume V and mean height h become fixed, and the terms in parenthesis become constant. Equation 3.10 may be simply written as

$$\nu = b.t \tag{3.11}$$

where $b = [\pi R^4.g.h/(8L.V)]$, a viscometer constant.

The viscometer constant can be either determined by obtaining values of the parameters needed or, more conveniently, by measuring the efflux time for a fluid of known kinematic viscosity, shown as follows.

$$b = \nu_{known}/t_{known} \tag{3.12}$$

Once the value of the viscometer constant is known, the kinematic viscosity of a test fluid (unknown) can be determined by measuring its efflux time and using equation 3.11. From a knowledge of the density, the viscosity, μ, is then readily calculated.

For efflux times less than 200 s, the capillary entrance and exit effects (called the end effects) become important and a modified version of equation 3.11 is used. (See Van Wazer, Lyons, Kim, and Colwell, 1963)

3.2.2. Rotational Viscometer

Rotational viscometers allow continuous measurements of the shear stress vs. shear rate relationship. The concentric-cylinder or coaxial-cylinder viscometer is one type of rotational viscometer commonly used in rheological studies (Fig. 3.3). In principle, the torque M, required to rotate the inside cylinder at a given number of revolutions per minute, is measured to obtain the flow curve for the material filled in the gap between the concentric cylinders.

The equation of motion for a coaxial-cylinder viscometer gives

$$\frac{1}{R^2}\frac{d}{dR}(R^2 T_{R\theta}) = 0 \tag{3.13}$$

or

$$R^2 T_{R\theta} = \text{constant} \tag{3.14}$$

where $T_{R\theta}$ is the shear stress exerted in the angular direction θ by a fluid surface at radius R on the fluid in the region $R_i < R < R_o$, R_i is the radius of the inner cylinder and R_o is the radius of the outer cylinder. Now the torque M exerted at the radial position R is given by

$$M = F.R \tag{3.15}$$

where F is the force exerted. From the definition of shear stress, $T_{R\theta}$, and substituting for the force from equation 3.15,

$$T_{R\theta} = \frac{F}{A} = \left(\frac{M}{R}\right)\left(\frac{1}{2\pi RL}\right) \tag{3.16}$$

where A is the surface area at radius R, and L is the length. Therefore,

$$R^2 T_{R\theta} = \frac{M}{2\pi L} \tag{3.17}$$

For a power-law fluid without yield stress, the shear-stress vs. shear-rate relationship in cylindrical coordinates is

$$T_{R\theta} = -m\left(R\frac{d\omega}{dR}\right)^n \tag{3.18}$$

where ω is the angular velocity (radians per unit time) of the inner cylinder. Substituting for $T_{R\theta}$ from equation 3.17 into equation 3.18, separating variables, and integrating between the radius of the inner cylinder and the radius of the outer cylinder gives the following expression:

$$-\int_{\omega}^{o} d\omega = (M/(2\pi mL))^{1/n} \int_{R_i}^{R_o} R^{-[(2+n)/n]} dR \tag{3.19}$$

Upon integration, equation 3.19 becomes

$$\omega = (M/(2\pi mL))^{1/n}(n/2)(R_i^{-2/n} - R_o^{-2/n}) \tag{3.20}$$

Taking the natural logarithm of equation 3.20 and rearranging it gives

$$\ln(M) = n\ \ln(\omega) + \ln(2\pi m.L) - n\ \ln(n/2) - n\ \ln(R_i^{-2/n} - R_o^{-2/n}) \tag{3.21}$$

Thus, rheological parameters m and n can be determined by plotting the experimental data in the form ln (M) versus ln (ω). The slope of the line of best fit gives the flow behavior index (n). The consistency coefficient m is obtained from the intercept.

In case of Newtonian fluids, n = 1 and m = μ. On substitution and rearrangement, equation 3.20 can be written as

$$\mu = (M/(4\pi\omega L))(1/R_i^2 - 1/R_o^2) \tag{3.22}$$

The foregoing expression is the Margules equation for Newtonian viscosity. Thus, a knowledge of torque M and angular velocity ω is all that is needed to determine Newtonian viscosity.

In a few viscometers the outer cylinder is a large container. In such systems it is reasonable to assume that $R_o >>> R_i$. On ignoring R_o, equations 3.20 and 3.22 for non-Newtonian fluids become

$$\omega^n = (M/(4\pi R_i^2 L))\,(2/m)(n/2)^n \tag{3.23}$$

and

$$\mu_a = (M/(4\pi\omega R_i^2 L)) \tag{3.24}$$

where μ_a is the apparent viscosity. Combining equations 3.23 and 3.24 yields

$$\mu_a = (2\omega)^{n-1}(m)(1/n)^n \tag{3.25}$$

Taking the natural log of equation 3.25 gives

$$\ln(\mu_a) = (n - 1)\ln(2\omega) + \ln(m) - n\ln(n) \tag{3.26}$$

Therefore, on plotting the measured values in the form of ln (μ_a) versus ln (2ω), a straight line is obtained. From the slope and intercept of the line of best fit, the flow behavior index n and the consistency coefficient m are determined.

3.2.3. Power Law Fluid with a Yield Value or Stress

When a yield stress τ_o exits for a fluid in a coaxial viscometer, the relationship between torque M and the rotor angular speed ω is given by (Charm, 1978):

$$\omega = \left(\frac{\tau_o}{m}\right)^{\frac{1}{n}} \int_{R_i}^{R_o} \left(1 - \frac{M}{2\pi\tau_o R^2 L}\right)^{\frac{1}{n}} \frac{dR}{R} \tag{3.27}$$

where

R_o = distance from the center at which the velocity of the streamline is zero. This will occur at the point in the gap where the yield stress is equal to the shear stress.

= radius of the outer cylinder in wide gap viscometer or equal to $[M/(2\pi\tau_o L)]^{1/2}$ for single-cylinder viscometer

R_i = radius of the cylindrical spindle

Since the rate of shear (γ) is proportional to the rotational speed of the viscometer, the τ_o may be determined by plotting $(M/L)^{0.5}$ versus $\sqrt{N}$. The intercept at $\sqrt{N} = 0$ is $(M/L)_o^{0.5}$ and

$$\tau_o = \left(\frac{M}{L}\right)_o \frac{1}{2\pi R_i^2} \tag{3.28}$$

By plotting log (N) versus log $[M/(2\pi\tau_o L.R_i^2)-1]$, the slope of the resulting line will be $1/n$. Then, using equation 3.27, m can be calculated.

3.2.4. Bingham Plastic Behavior Fluids

The relationship for Bingham plastics is given by Heldman and Singh(1981). When using coaxial rotational viscometer it becomes

$$\omega = \frac{M}{m.4\pi.L}\left(\frac{1}{R_i^2} - \frac{1}{R_o^2}\right) - \frac{\tau_o}{m}\ln\left(R_o/R_i\right) \tag{3.29}$$

Note: There are two major points worth noting.

1. The foregoing development does not account for the drag on the top and bottom of the rotating inner cylinder (called the end effects). To overcome this, the actual length of the inner cylinder is replaced by an effective length $L' > L$. Methods to determine L' are available in the literature (Van Wazer et al., 1963). Manufacturer's of these viscometers also provide such data for their system.

2. As with capillary viscometers, the assumptions made in the derivation of equations for the calculation of rheological parameters include: steady-state and laminar flow, time independent properties, no-slip condition at the wall, incompressible fluid, and isothermal conditions at measurement. These conditions must be strictly adhered to for the results to be meaningful.

3.3. OBJECTIVES

The specific objectives of this laboratory are to:

1. Determine the Newtonian viscosity of a test fluid using a capillary viscometer.

2. Determine the flow curves for Newtonian, pseudoplastic, and dilatant materials using a coaxial-cylinder viscometer.

3. Determine the effect of temperature on Newtonian viscosity.

3.4. APPARATUS

1. Capillary viscometer (Cannon--Fenske, size no. 100 (kinematic viscosity range 2 to 10 centistokes, cS) or size no. 150 (kinematic viscosity range 6 to 30 cS)
   Constant temperature glass water bath (10° to 100°C) (preferably Cannon M1 Series)
   Suction type rubber bulb, two 100 mL cylinders, two 100 mL volumetric flasks, stopwatch, balance
   Distilled water, acetone, and trichloroethylene
   Viscosity standards: silicone oils, water
   Test fluids: clarified apple juice, Gatorade, evaporated milk, or sucrose--water solutions with viscosity within the viscometer range

2. Brookfield Synchro-Lectric, Model LV or RV, or other similar coaxial-cylinder viscometer
   Constant-temperature bath, thermometer, and beakers (600 mL, A)
   Suggested test fluids:
   Newtonian: corn syrup, corn oil, STP oil additive
   Pseudoplastic: mustard paste, French salad dressing, Gerber bananas
   Dilatant: cornstarch in water, 55% by weight

3. Brookfield Synchro-Lectric, RV or LV model, or other similar viscometer
   Two constant temperature water baths (30° to 80°C), thermometers, beakers (600 mL, one per bath)
   Test fluid: corn syrup or corn oil

3.5. PROCEDURE

3.5.1. Determination of the Newtonian Viscosity Using a Capillary Viscometer

The Newtonian viscosity data for this part of the experiment will be obtained with the Cannon--Fenske Capillary Viscometer, shown in Fig. 3.2, size no. 100. Larger size may be used for higher- viscosity fluids.

1. Fill up a 10-mL graduated cylinder with the test fluid.

2. Attach a suction bulb to arm *G* of the viscometer. Invert the viscometer, dip arm *A* in the test fluid, and apply suction until the fluid level reaches etched mark *E*. Return the viscometer to an upright position.

3. Place the viscometer in a controlled temperature glass water bath and allow the temperatures to equilibrate. Record the temperature.

4. Record the efflux time θ for the test fluid to drain between marks *C* and *E*.

5. Repeat the measurement by applying suction to arm *A* to bring the test fluid level above mark *C*.

6. Rinse the viscometer thoroughly, first with distilled water and then with acetone. Dry the viscometer completely.

7. Follow steps 1 to 5 for other test fluid or the viscosity standards.

8. Clean the viscometer after each standard first with trichloroethylene and then with acetone. Aspirate to dryness.

9. Determine the densities of the test fluid and the standard by filling them in tared 100 mL volumetric flasks and weighing. Record the data in Table 3.1.

3.5.2. Determination of Rheological Parameters Using a Coaxial-Cylinder Viscometer

The rheological parameters are determined from a knowledge of the shear-stress vs. shear-rate relationship generated with a coaxial-cylinder viscometer, such as the Brookfield Synchro-Lectric model RVT or Haake Rotovisco RV series. In such types of viscometers a spindle rotates in the fluid at a selected angular velocity and the torque required to overcome the viscous resistance of the fluid is measured. The rheological parameters are computed from these measurements at various angular velocities, discussed earlier.

The following procedure is for obtaining the required data with a Brookfield Synchro-Lectric model RVT, recommended by the manufacturer.

1. Record the product (test fluid) information on the data sheet B of Table 3.1 along with the instrument data.

2. Pour about 500 mL of the test fluid into a 600-mL beaker and place the beaker in a temperature-controlled water bath (30°C). Record temperature of the fluid after equilibration.

3. Attach spindle to viscometer shaft, taking care to avoid exerting side thrust on the shaft.

4. Insert spindle in the test fluid up to immersion groove cut in the spindle shaft.

5. Level the viscometer by adjusting screws on mounting stand and the bubble level on the dial casing.

6. Depress the clutch and start the motor. Release the clutch and allow the dial to rotate until the pointer stabilizes at a position on the dial.

7. Depress the clutch to freeze the pointer. Stop the motor and record the rotational speed and the dial reading (torque M). If the dial reading is less than 30, change to a larger spindle (lower number). For off-scale dial reading, a smaller spindle (higher number) should be used.

8. Stop motor and record the dial reading and the rotational speed.

9. Reset the rotational speed (angular velocity) and repeat steps 6 through 8. For good results, measure torque (dial reading) at four or more angular velocities.

10. Obtain data on at least one Newtonian, one pseudoplastic, and one dilatant and record them in Table 3.1(B).

3.5.3. Determination of Effect of Temperature on Newtonian Viscosity

In addition to the data on the Newtonian viscosity of the test fluid obtained in the previous experiment using a coaxial- cylinder viscometer, two more data sets can be obtained by conducting measurements at two more temperatures (50° and 70°C). The procedure to be followed is essentially as described for the previous experiment.

Although it is possible to determine the Newtonian viscosity from a single measurement of shear rate (angular velocity) and shear stress (torque), measurements at several velocities should be made to show that the flow behavior index n remains unity at all temperatures of measurement.

<u>Notes</u>

1. To convert revolutions per minute (rpm) to radians per second, use the following formula:

$$\frac{\text{revolutions}}{\text{min}} \; x \; \frac{\text{min}}{60 \text{ sec}} \; x \; \frac{2\pi \; radian}{revolution} = \omega$$

To convert dial reading to torque M (dyne.cm), use

$$\frac{\text{dial reading}}{100} \; x \; Factor = torque \; (M), \quad dyne.cm$$

The factor is supplied by the viscometer manufacturer. For example, for the Brookfield Synchrolectric Model RVT, the factor is 673.7.

2. The following data are supplied by Brookfield Engineering Laboratories for their cylindrical spindles.

Spindle No.	R(cm)	L'(cm)	L(cm)
LV-1	0.9421	7.493	6.510
LV-2	0.5128	6.121	5.395
LV-3	0.2941	4.846	4.287
LV-4	0.1588	3.396	3.101
LV-5	0.1588	1.684	1.514

3.6. RESULTS AND DISCUSSION

1. Report the capillary viscometer constant b and the viscosity of one Newtonian test fluid.

2. Plot the flow curves (torque versus angular velocity) for one Newtonian, one pseudoplastic, and one dilatant test fluid using the data obtained with the coaxial-cylinder viscometer.

3. Determine m and n for the non-Newtonian test fluids using plots of ln (M) against ln (ω). Write the specific power law for each of test fluid.

4. Determine the viscosity of the Newtonian test fluid at three temperatures and compute Arrhenius parameters by plotting ln (M) versus $1/T$.

5. A Cannon--Fenske capillary viscometer is to be calibrated. Two standards with known viscosity and density are tested. The following data were obtained:

Sample	Viscosity, Pa.s	Density, kg/m^3	Efflux Time, s
Standard 1	$1.2(10^{-3})$	$1.01(10^3)$	59.4
Standard 2	$1.3(10^{-3})$	$1.03(10^3)$	63.1

The viscometer is calibrated by finding b, the viscometer constant. From the foregoing data, find b for this particular viscometer.

6. Flow behavior index n and consistency coefficient m of banana puree at various temperatures are listed as follows.

Temperature, °C	m, Pa.s^n	n
19	7.0	0.458
24	6.5	0.458
53	3.5	0.458
74	2.9	0.459

Discuss the effect of temperature on *m* and *n*.

Note: The computer program "Rheology" can be used for most of these calculations.

Table 3.1 DATA SHEET FOR RHEOLOGY LAB

DATE: ______________________ GROUP: ______________

A. Newtonian Fluids (capillary viscometer)

Instrument data: Manufacturer ______________________
Serial No. ______________ Model No. ______________

Product data:
Product 1 ______________
Product manufacturing and label information ______________________

Product 2 ______________
Product manufacturing and label information ______________________

Product 3 ______________
Product manufacturing and label information ______________________

Viscometer

Product	Density, g/mL	Viscosity, cp	Efflux Time, s
			1. 2. 3.
			1. 2. 3.
			1. 2. 3.

B. Power Law Fluids (Brookfield or similar viscometer)

Instrument data: Manufacturer ______________________
Serial No. ______________________ Model No. ______________

Product data:
Product 1 ______________________
Product manufacturing and label information ______________________
____

Product 2 ______________________
Product manufacturing and label information ______________________
____

Product 3 ______________________
Product manufacturing and label information ______________________
____

Product 4 ______________________
Product manufacturing and label information ______________________
____

Viscometer

	Torque M (dial reading)			
Rotational speed, rpm	Product 1 (T = °C)	Product 2 (T = °C)	Product 3 (T = °C)	Product 4 (T = °C)
Spindle no.				
Spindle constant				

C. Power Law Fluids (Haake or similar viscometer)

Instrument data: Manufacturer ______________________________
Serial No. ______________________ Model No.______________________

Product data:
Product ______________________________
Product manufacturing and label information ______________________________

Viscometer	
Rotational Speed	Torque M (dial reading)

D. Viscosity-Temperature Relationship (Brookfield or similar viscometer)

Instrument data: Manufacturer ______________________________
Serial No. ______________________ Model No.______________________

Product data:
Product ______________________________
Product manufacturing and label information ______________________________

	Torque M (Dial Reading)			
Rotational speed, rpm	(T = °C)	(T = °C)	(T = °C)	(T = °C)
Spindle no.				
Spindle constant				

Prelab Questions for Laboratory 3

NAME:__________________________________ DATE:____________________

ANSWER PRELAB QUESTIONS ON THIS SHEET

Laboratory 4

EVALUATION OF A PUMP PERFORMANCE

SUMMARY

Pumps are a major unit operation equipment used in the process industry to transport fluids from one place to another. This laboratory is designed to acquaint you with the types of pumps, operation of pumps, terminologies used, guidelines for the selection of pumps, and determination of the characteristic curves. Characteristic curves are relationships between the differential head, pumping volume, speed, and efficiency of a particular pump and are useful in pump selection and operation.

4.1. BACKGROUND

4.1.1. Pumps

A pump is a device that adds energy to the system on which it is operating. This energy can be used to increase the pressure (head) in the discharge line of a pump or it can be used to move a fluid through the discharge lines. Usually part of the energy is used for each of these purposes.

Pumps are widely used in many industries such as those processing food, petrochemical, and petroleum-related products. In the food processing industry, the efficient operation of refrigeration systems, homogenizers, high temperature-short-time pasteurizers, and many unit operations are partially dependent on the proper use of pumps.

4.1.2. Developed Head

A typical pump application is shown diagrammatically in Fig. 4.1. The pump is installed in a pipeline to provide the energy needed to draw liquid from a reservoir and discharge a constant volumetric flow rate at the exit of the pipeline Z_b, above the level of the liquid. At the pump itself, the liquid enters the suction connection at station a and leaves the discharge connection at station b. An energy balance, using the Bernoulli equation, can be written between stations *a* and *b*.

Total energy at b = total energy at a + actual energy added to the system by pump

A fraction of energy added by the pumps gets dissipated by frictional forces inside the casing and impeller. If a pump adds W amount of mechanical energy at an efficiency η, a net of $\eta.W$ is the quantity of energy added to the fluid. The total energy at points a and b would be made of pressure and kinetic and potential energies. If we neglect the losses in the lines outside the pump and do a mechanical energy balance between points a and b, we get (neglecting other energies such as thermal, magnetic, etc.):

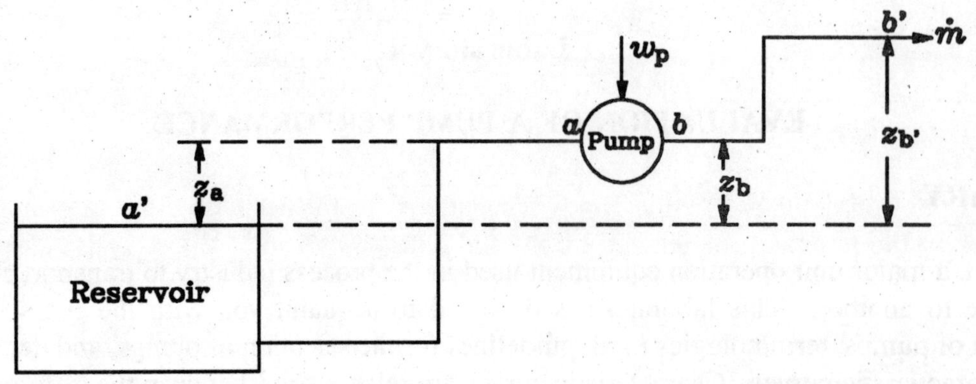

Fig. 4.1 **A typical pump flow system.**

$$W_p = \left(\frac{p_b}{\rho} + \frac{g \cdot Z_b}{g_c} + \frac{\alpha \cdot V_b^2}{2g_c}\right) - \left(\frac{p_a}{\rho} + \frac{g \cdot Z_a}{g_c} + \frac{\alpha \cdot V_a^2}{2g_c}\right) \qquad (4.1)$$

where p is pressure, Z is elevation above a reference level, V is average fluid velocity, ρ is fluid density, g is acceleration due to gravity, W_p is work input to pump, α is 1.0 for turbulent flow and 2.0 for laminar flow in a circular conduit, and $g_c = 1$. The quantities in the parentheses are called total heads and are denoted by H, or

$$H = \frac{p}{\rho} + \frac{g \cdot Z}{g_c} + \frac{\alpha \cdot V^2}{2g_c} \qquad (4.2)$$

In pumps, usually the difference between the heights of the suction and discharge connections is negligible, and Z_a and Z_b can be dropped from equation 4.1. If H_a is the total suction head, H_b the total discharge head, and $\Delta H = H_b - H_a$. Equation 4.1 can be written as

$$W_p = \frac{H_b - H_a}{\eta} = \frac{\Delta H}{\eta} \qquad (4.3)$$

where η is pump efficiency.

4.1.3. Power Requirement

The power supplied to the pump drive from an external source is denoted by P'_b. It is calculated from W_p by the following expressions:

$$P'_b = \dot{m} . W_p = \frac{\dot{m} . \Delta H}{\eta} \qquad (4.4)$$

where $\dot{m}$ is the mass flow rate.

The power delivered to the fluid is calculated from the mass flow rate and the head developed by the pump. It is denoted by P'_f and defined by

$$P'_f = \dot{m}\, \Delta H \qquad (4.5)$$

From equations 4.4 and 4.5,

$$\frac{P'_f}{P'_b} = \eta \qquad (4.6)$$

Power is delivered to the pump by using a prime mover (an electrical motor or an engine), thus the efficiency of the prime mover is also included in determining actual horse power of the prime mover necessary to run a pump to discharge a specific quantity of fluid.

4.1.4. Suction Lift and Cavitation

The power calculated by equation 4.3 depends on the difference in pressure between discharge and suction. From energy considerations it is immaterial whether suction and discharge pressures are below atmospheric pressure or well above it. Practically, the lower limit of the suction pressure is fixed by the vapor pressure of the liquid corresponding to the temperature of the liquid at the suction. If the pressure on the liquid reaches the vapor pressure, some of the liquid flashes into vapor, a process called cavitation. When cavitation occurs in the suction line, no liquid can be drawn into the pump. Cavitation will not occur if the sum of the velocity and pressure heads at the suction is sufficiently greater than the vapor pressure of the liquid. The excess of the sum of these heads over the vapor pressure is called the net positive suction head (NPSH), denoted by H_{sv}. For a pump taking suction from a reservoir like that shown in Fig. 4.1 the NPSH is:

$$H_{sv} = \frac{\alpha . V_a^2}{2g_c} + \frac{p_a - p_v}{\rho} \qquad (4.7)$$

where p_v is the vapor pressure. An energy balance written between station *a'*, at the level of the liquid in the reservoir, and station *a*, at the suction of the pump, gives, considering $Z_a' = 0$ and $V_a' = 0$,

$$\frac{p'_a}{\rho} = \frac{p_a}{\rho} + \frac{\alpha . V_a^2}{2g_c} + h_{fs} + \frac{g . Z_a}{g_c} \qquad (4.8)$$

where h_{fs} is the friction in the suction line. Eliminating $p_a/\rho + \alpha.V_a^2/2g_c$ from equations 4.7 and 4.8 gives

$$H_{sv} = \frac{p'_a - p_v}{\rho} - h_{fs} - \frac{g . Z_a}{g_c} \qquad (4.9)$$

For the special situation where the liquid is nonvolatile ($p_v = 0$), the friction is negligible ($h_{fs} = 0$), and the pressure at station *a*' is atmospheric, the NPSH is the barometric head as measured by a column of the liquid and represents the maximum possible suction lift from a tank vented to the atmosphere. For cold water this is about 34 ft or 10 m.

Cavitation may occur within a pump if the pressure at any point reaches that corresponding to $H_{sv} = 0$. This not only prevents normal pump operation but causes severe erosion and mechanical damage. The NPSH must therefore be greater than zero and usually is 2 or 3 m or more. It is the responsibility of the engineer who designs the process to state the design value of H_{sv}, and it is the responsibility of the pump supplier to ensure that the pump operates properly at this condition.

4.1.5. Types of Pumps

Pumps are classified into two groups: variable head capacity type (kinetic) and positive displacement type, and their selection is based on service required, type of fluid to be pumped, discharge pressure necessary, viscosity of fluid, temperature of fluid, and the availability of net pressure suction head (NPSH). In case of variable head-capacity type pumps, the discharge pressure varies as the capacity of the pump is regulated, whereas the latter type has constant discharge pressure at different flow rates. Regulation of flow rate in positive displacement pumps is done by changing the displacement rate or capacity of the intake chamber.

1. Variable head capacity pumps (kinetic)
   a. Radial flow (centrifugal)
   b. Axial flow (propeller)
   c. Mixed flow (turbine)

2. Positive displacement pumps
   a. Reciprocating pumps
   b. Rotary pumps

Centrifugal Pumps. Centrifugal pumps are used in high capacity and low head services. In this pump, the impeller utilizes centrifugal forces to impart pressure to the fluid. Centrifugal pumps are used for most of the refinery services, chemical industry, and food industry and can handle

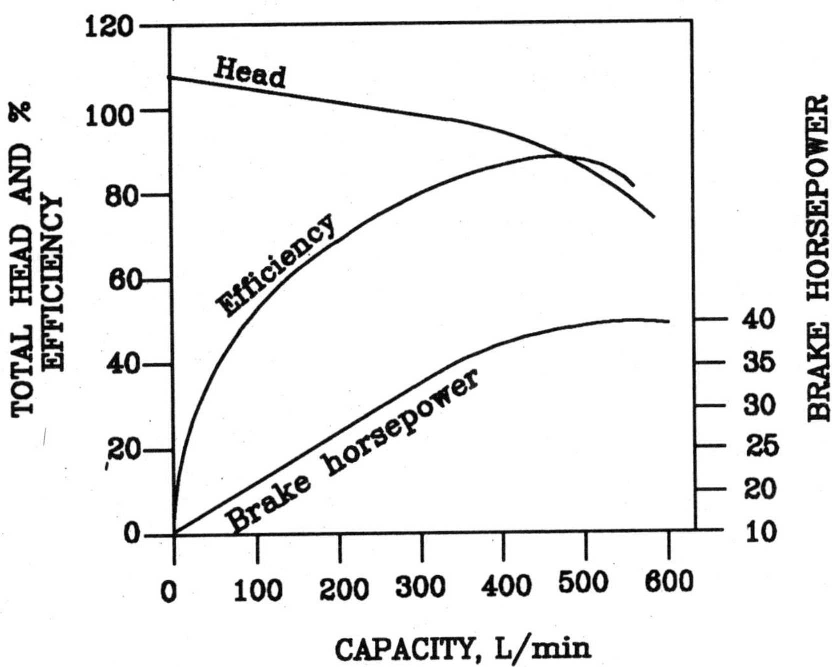

Fig. 4.2 **Typical characteristic curves for a centrifugal pump.**

liquids containing solids, grit, and corrosive fluids. Fluids with kinematic viscosity up to 650 cS can be handled, although 440 cS is considered the practical limitation. At a fixed impeller speed this pump develops the same head regardless of the density of the fluid. Figure 4.2 shows typical characteristic curves, after plotting total head, percent efficiency, and brake horsepower (BHP) against the flow rate of fluid.

Centrifugal pumps can be built with a single stage or multiple stages to boost the discharge pressure. Horizontal pumps are most commonly used in routine usages. Centrifugal pumps need very low maintenance. An important requirement for this type of pump is that there should be sufficient net pressure suction head (NPSH) at suction above the vapor pressure of the fluid handled to prevent vaporization within pump. Above 2 to 3 m NPSH is generally acceptable in the industry. Low NPSH pumps are more expensive. The NPSH requirement for boiling liquids is more critical. It is possible to get pumps with NPSH requirements as low as 1.5 m; this usually involves the installation of an oversize pump. For cases where available NPSH is less than 1.5 m, vertical pumps can be obtained. In this case, NPSH can be reduced to zero. But their construction is usually considerably more expensive than the horizontally mounted pumps.

In general, a centrifugal pump should not be run continuously at less than 30% of its nominal capacity. Nominal capacity in this case is defined as the midpoint of the capacity range in which it would normally be operated.

If continuous operations will be below the minimum rating indicated, then the fluid pumped must be recirculated to maintain at least the minimum flow but preferably somewhat more. There is no power loss in recirculation because of low efficiency at low flow rate and high efficiencies at nominal rating. Normally, it is desirable to recirculate through the drum or storage to provide adequate surface for radiation of the heat put into the fluid by the pump. Recirculation should be at least 1.0 m^3/h or more to permit the use of a reasonable size of recirculation orifice or control valve.

The following are approximate relationships of capacity, head, brake horsepower (BHP), with pump speed N at constant impeller diameter.

Capacity (Q): $$\frac{Q_1}{Q_2} = \frac{N_1}{N_2} \tag{4.10}$$

Head (H): $$\frac{H_1}{H_2} = \frac{N_1^2}{N_2^2} \tag{4.11}$$

BHP: $$\frac{\text{BHP}_1}{\text{BHP}_2} = \frac{N_1^3}{N_2^3} \tag{4.12}$$

or

$$\frac{Q_1}{Q_2} = \frac{N_1}{N_2} = \left(\frac{H_1}{H_2}\right)^{1/2} = \left(\frac{BHP_1}{BHP_2}\right)^{1/3}$$

The main advantages of centrifugal pumps are that they:

1. Operate at speeds that allow direct connection to steam turbines and electric motors.

2. Are compact and low cost in larger sizes.

3. Have smooth flow through the pump and uniform pressure in the discharge pipe.

4. Have low maintenance cost and repair times.

5. Have power characteristics that make it an easy load for its driver. An increase in head reduces the power requirement, a characteristic that makes overloading of the motor by closing the discharger impossible.

Positive Displacement Pumps. In the positive displacement pumps, a definite volume of liquid is trapped in a chamber that is alternately filled from the inlet and emptied at a higher pressure through the discharge. There are two subclasses of positive displacement pumps. In reciprocating pumps the chamber is a stationary cylinder that contains a piston or plunger; in rotary pumps the chamber moves from inlet to discharge and back to the inlet.

Reciprocating Pumps. In a reciprocating pump, liquid is drawn through an inlet check valve into the cylinder by the withdrawal of a piston and then forced out through a discharge check valve on the return stroke. Most piston pumps are double-acting (i.e., liquid is admitted alternately on each side of the piston so that one part of the cylinder is being filled while the other is being emptied). Two or more cylinders are often used in parallel with common suction and discharge headers, and the configuration of the pistons is adjusted to minimize fluctuations in the discharge rate. The piston may be motor driven through reducing gears or a steam cylinder may be used to drive the piston rod directly. The maximum discharge pressure for commercial pistons is about 50 atm (5065 kPa).

Plunger pumps are used for higher pressures. A heavy-walled cylinder of small diameter contains a close-fitting reciprocating plunger, which is merely an extension of the piston rod. At the limit of its stroke the plunger fills nearly all the space in the cylinder. Plunger pumps are single acting and usually are motor- driven. They can discharge against a pressure of 1500 atm (152 MPa) or more.

The mechanical efficiency of reciprocating pumps varies from 40 to 50% for small pumps to 70 to 90% for large ones. It is nearly independent of speed within normal operating limits and decreases slightly with increasing discharge pressure because of added friction and leakage.

The ratio of the volume of fluid discharged to the volume swept by the piston or plunger is called the volumetric efficiency. In positive displacement pumps the volumetric efficiency is nearly constant with increasing discharge pressure, although it drops a little because of leakage. For a reciprocating pump it varies from about 90 to 100%.

Reciprocating pumps are used in special services. Special circumstances that may favor use of reciprocating pumps include the following:

1. Where equipment vendors make a few centrifugal pumps of low NPSH

2. Rebuilding jobs where large numbers of used reciprocating pumps are available

3. Sludge and slurry service

4. Proportionating pumps

5. Relatively low capacities (0 to 5.0 m^3/h) at high discharge heads where a centrifugal pump with recirculation is not practical

6. Heads above 1500 m

7. Viscous fluid such as asphalts and greases at high temperatures

8. Intermittent services such as pump out where wide range of capacities and fluids must be handled and where cost is lower than centrifugal

9. Where pulsating flow can be tolerated

Rotary Pumps. A wide variety of rotary positive displacement pumps are available, such as gear, lobe, screw, cone, and vane. Unlike reciprocating pumps, rotary pumps contain no check valves. Rotary pumps operate best on clean, moderately viscous fluids (e.g., light lubricating oil). Discharge pressures up to 200 atm (20 MPa) or more can be attained as they are dependent on close clearance for dependable operation. These pumps usually are at a disadvantage for dirty or abrasive liquids, especially when high service factor is required. Because rotary pumps have a nonpulsating flow they have excellent process characteristics in their range of operation.

The chief advantage of rotary pumps over reciprocating pumps are that rotary pumps:

1. Deliver a continuous flow that is practically free from pulsations. Large air chambers to absorb the shock of intermittent discharge are not required.

2. Are similar in construction. There are no valves to be opened and closed with each successive quantity delivered.

3. Are smaller in dimension for a given capacity, therefore they occupy less space.

4. Are easier to install. They do not require large foundations to absorb the shocks of reciprocating parts and intermittent delivery.

5. Are easier to maintain. There are no valves or valve springs to be serviced periodically or replaced.

4.1.6. Performance of Centrifugal Pumps

The developed head of a pump is considerably less than that calculated from the ideal pump relation. Also, efficiency is less than unity and the fluid horsepower is less than the ideal horsepower. Different sources of head losses include: loss of head, circulatory flow; fluid friction; and shock loss.

Power Losses. Fluid friction and shock losses lead to the losses of power as they convert of mechanical energy into heat. Three other kinds of loss that do not cause a loss in head but do increase power consumption occur in a centrifugal pump are leakage, disc friction, and bearing losses. The barrier to a reverse flow provided by the wearing ring is not perfect, and a certain amount of unavoidable interior leakage occurs from the discharge of the impeller back to the suction eye. The effect of leakage is to reduce the volume of actual discharge from the pump per unit of power expended. The extra work used to maintain the leakage is converted into heat and lost.

Disk friction is the friction that occurs between the outer surface of the impeller and the inside of the casing. It includes the greater friction caused by the pumping action on the liquid that occurs in that same space. Liquid in contact with the rotating impeller is picked up and thrown outward toward the volute. The liquid then flows back along the inside wall of the casing to the shaft, again to be picked up by the impeller and repumped. Power is required to maintain this useless secondary action. Bearing losses constitute the power required to overcome mechanical friction in these bearing and stuffing boxes of the pump.

Efficiency. The efficiency is the ratio of the delivered power to the total power input. The efficiency is maximum at the design flow rate and decreases at other rates, largely because of the variation of shock loss with flow rate.

4.1.7. Pump Priming

The theoretical head developed by a centrifugal pump depends on the impeller speed, the radius of the impeller, and the velocity of the fluid leaving the impeller. If these factors are constant, the developed head is the same for fluids of all densities and is the same for liquids and gases. The increase in pressure, however, is the product of the developed head and the fluid density. If a pump develops, say a head of 30 m, and if the pump is full of water, then the increase in pressure is (30 m) (1000 kg/m^3) (9.81 m/s^2) = 294 kPa. If the pump is full of air at normal density (25 °C, 1 atm. pressure), the pressure increase is about 350 Pa. A centrifugal pump trying to operate on air, then, can neither draw liquid upward from an initially empty suction line nor force liquid along a full discharge line. A pump with air in its casing is airbound and can accomplish nothing until the air has been replaced by a liquid. Air can be displaced by priming the pump from an auxiliary priming tank connected to the suction line or by drawing liquid into the suction line by an independent source of vacuum. Also, several types of self-priming pumps are available. Positive displacement pumps can compress a gas to a required discharge pressure and are not subjected to air binding.

Prelab Questions

Q1. Why is the NPSH of a centrifugal pump so important?

Q2. Which pump will be more expensive? (a) to pump sewage water or (b) to pump boiling water at atmospheric pressure. Why?

Q3. How will you regulate the flow rate in centrifugal pump? Would you regulate it by placing a valve on the suction side of pump?

Q4. Power to the pump is supplied by running an electric motor or steam turbine. What will you recommend to the food industry? why?

Q5. What is a shutoff pressure of a centrifugal pump?

Q6. What is the difference in operating principles of centrifugal and reciprocating pumps?

Q7. Why is available NPSH not so critical in selection of positive displacement pumps?

Q8. How do you regulate the flow rate in positive displacement pumps? What is this difference in regulator controls?

Q9. Why is it not recommended to place any ordinary or control valve on discharge side of positive displacement pumps?

Q10. As you may notice, the piston-type positive displacement pumps have a pulsating discharge flow. Why does this pulsation occur and how can you minimize it?

Q11. What is the difference between the NPSH and priming of centrifugal pumps?

Q12. Why is the priming and NPSH not so critical for the reciprocating type of pumps?

Q13. Why is it not dangerous to close the valve on the discharge side of a centrifugal pump when it may be fatal to do so in the case of positive displacement-type pumps?

4.1.8. Pump Performance--Centrifugal Pumps

If a valve in the discharge line of a centrifugal pump is closed, no liquid fluid will flow and all of the energy delivered by the pump goes to increase the pressure in the discharge line. If there is no restriction in the discharge line (valve fully open), all of the available energy will move the fluid and the maximum flow is obtained. Between these two extremes there are various combinations of flow and pressure possible for a centrifugal pump. A graph that relates the pressure (head), efficiency, and input power with the flow of a centrifugal pump is called its characteristic curves. The pump characteristic will depend on such properties of a pump as its speed and the diameter, width, and design of the impeller. The characteristic of a pump is very important for evaluating its performance and efficiency. Figure 4.2 shows a typical characteristic curve of a centrifugal pump.

4.1.9. Calculations

As shown in Fig. 4.2, a characteristic curve is obtained by plotting head, efficiency, and brake horsepower versus flow rate. The head is the differential pressure in terms of the equivalent height of fluid being pumped.

1. The differential pressure measured (p) across a pump can be converted into the fluid head (H) by

$$H = \frac{p}{\rho} + \frac{g.Z}{g_c} + \frac{\alpha . V^2}{2g_c} \qquad (4.2)$$

2. You will be measuring the power consumed by the system using a wattmeter. You should be able to find the BHP of the pump. How would you do it?

$$\text{Normal power consumption (kW)} = \frac{\text{BHP } (P'_B) \text{ X } 746/1000}{\text{motor efficiency (in decimals)}}$$

3. To plot efficiency vs. flow rate, what information would you need?

P'_f = theoretical power (hydraulic horsepower) needed = $\Delta H.\dot{m}\ g.\rho\ /\ g_c$, W
where

ΔH	=	differential head, m
$\dot{m}$	=	fluid flow rate, m^3/s

4. η = efficiency of the pump = $P'_f\ /\ P'_B$

4.2. OBJECTIVES

The main objectives of this lab are to:

1. Acquaint you with the functions, applications, and operation of various types of pumps used in food industry.

2. Provide general guidelines for the selection of a pump for a given situation and to make you aware of several technical terminologies often used in process industry relating to various pumps.

3. Demonstrate the differences between positive displacement pumps and centrifugal pumps.

4. Determine the characteristic curve of a centrifugal pump.

4.3. APPARATUS

1. A sanitary centrifugal pump.

2. A "constant" - level tank on the suction side of the pump.

3. A means for measuring flow. This might include:
   (a) A scale, a weight tank with a bypass mass circuit, a three-way valve, and an electric timer.
   (b) A sanitary volumetric (lobe-type positive displacement) meter and an electronic timer.
   (c) A glass or transmitting rotameter.
   (d) A weir, float or height gauge, and discharge tank.
   (e) An orifice meter, Venturi meter, or nozzle.
   (f) A wobbling (also called "nutating") disc "water" meter.
   (g) A turbine meter with pulse pickup, pulse counter, and electronic differentiating device.
   (h) A Thompson meter.
   (i) A discharge tank of known dimensions, a depth gauge, and a timer.

   Approaches (a), (b), (c), and (i) are preferred.

4. A safe means of measuring electric power. This requires grounding the pump and meter separately as well as ensuring that the ground wire is not electrically interrupted by the installation of the measuring circuit and shielding the instrumentation and connections from splashes. HIGH VOLTAGE AND WATER MEAN A NEED FOR CARE. When these materials are handled safely, there is no need for alarm.

   Equipment may include:
   (a) A watt-hour meter and a timer
   (b) A three-phase watt-meter, for a three-phase motor
   (c) Two single-phase watt-meters, for a three-phase motor
   (d) A single-phase watt-meter for a single-phase motor

   A useful device is an enclosed, watertight, grounded electrical box with an inlet cord and plug, a grounded outlet, labeled screw lugs and jumpers for voltage and current meters, and a water proof transparent cover. The watt-meter connections should also be protected from water spray.

5. A means of measuring discharge head or pressure which may include one of the following:
   (a) A sanitary pressure gauge with a small least-division.
   (b) A regular pressure gauge with a small least-division, a pressure snuffer to reduce measurement noise due to turbulence, a means of backflushing the gauge (must have drain opening at the end of the border tube), and a snuffer.
   (c) A mercury manometer with calming reservoir after a restriction on the inlet and a discharge surge trap for blown mercury.
   (d) A diaphragm-type pneumatic differential pressure gauge and precision test gauge readout or a pressure-to- electric converter and readout.
   (e) A piezoelectric or diaphragm-type circular strain gauge cell and appropriate read-outs.

Approach (a) is preferred for students, but students will have to determine the mean pressure by the "method of swings" used for the manual mass balances.

6. A means of measuring the suction pressure. This may be any of the devices outlined in part 5.

7. A means of measuring the physical dimensions of the system, especially the difference between the center of the pressure sensors and the center of the pump, and the feed level, and the center of the tank.

8. A means of adjusting flow rate such as a metering valve in the discharge line or back pressure valve. A plug valve is generally not satisfactory due to the exceptionally large change in flow area for a small change in valve position as the valve is initially opened.

9. A means of measuring product density such as a hydrometer, "westphal" balance, or pycnometer.

INSTRUCTIONS FOR THE STUDENT

To have this exercise be fun, you need to be prepared to:

1. Convert pressure to head by using the Bernoulli equation.
2. Convert elevation to head.
3. Convert watt-hours and time to watts and use the K_h of the meter for watt-hours per turn (correct determination of power consumption).
4. Calculate hydraulic horsepower from the fluid head and mass flow rate.
5. Convert horsepower to watts.
6. Calculate efficiency (i.e., (output power/input power) x 100).

4.4. PROCEDURE

1. Acquaint yourself with the functions, applications, and operations of various types of pumps.

2. Make a visual inspection of a positive displacement and centrifugal pump. Find out the difference in construction.

3. Select a centrifugal pump to determine its characteristic curves.

4. Connect all the equipment as described in "APPARATUS" section necessary to run the experiment properly.

5. Select a pump speed and measure the power consumption, flow rate, suction pressure, and discharge pressure of the pump. Record the data in Table 4.1.

6. Change the speed of the pump and repeat the experiment by monitoring power consumption, flow rate, and suction and discharge pressures.

7. Measure fluid temperature, viscosity, and density.

8. Do necessary calculations and record in Tables 4.2 and 4.3. Computer program "Pumps" as explained in the "Computer Programs" Chapter can be used for these calculations.

NOTE: The data should be collected for at least six pump speeds to get proper characteristics curves.

4.5. RESULTS AND DISCUSSION

1. Draw a schematic diagram of the setup.

2. Plot the head (meter of water), power input, and pump efficiency vs. flow rate (m^3/h) for each of the pump speeds.

3. Discuss the pump characteristics. How you will use these plots in the selection of a pump for a particular application?

4. What is the effect of pump speed on the characteristic of a centrifugal pump? Why?

5. When the valve of the discharge line of a centrifugal pump is fully closed, energy is still being supplied to the system by the pump. What happens to this energy under these conditions?

6. Theoretically compare the performance of the positive displacement and the centrifugal pump. Discuss the difference and provide some explanation for these differences.

7. What is cavitation? How can it be minimized?

8. List potential experimental problems and methods to improve them.

4.6. REQUIREMENTS FOR REPORT

1. Raw data: see Tables 4.1 to 4.3

2. Sample calculations

3. Four graphs.
   (a) Differential head (ft or m) and efficiency vs. revolutions per minute with valve full open and corresponding flow rate (ft^3/min or m^3/s)
   (b) Differential head (ft or m) and efficiency vs. flow rate with valve full, half, and a third open for three revolutions per minute.

4. In the discussion section, state the calculated power factor (PF) and how it could be used.

Note: *Power Factor (PF).* An alternative method of calculating watts utilizes the power factor. This method may be used when a watt-meter is unavailable. Yet, a watt-meter must be made available to determine the power factor. The power factor is a dimensionless constant that is unique for a given system. It is the ratio between actual power measured by a watt-meter and apparent power (volts x amperes in single phase). The power factor multiplied by current at a given RPM, voltage, and the number of phases (three here) of the motor yields watts.

$$\text{Watts} = 3 \times \text{Volts} \times \text{Amps} \times \text{PF}$$

4.7. PROBLEMS ON A COMPUTER

1. Calculate and compare (a) the developed head of the pump, (b) the total power input, and (c) the net positive suction head (NPSH) for a centrifugal pump transporting water, benzene, ethyl alcohol, and milk. The reservoir is at the atmospheric pressure. The gauge pressure at the end of the discharge line is 50 $lb_f/in.^2$ (345 kN/m^2) gauge. The discharge is 32.8 ft (10 m), the pump suction is 4.9 ft (1.5 m) above the level in the reservoir. The discharge line is 1.5 in. (3.81 cm) inside diameter. The friction in the suction line is known to be 0.5 $lb_f/in.^2$ (3.45 kPa) and that in the discharge line is 5.5 $lb_f/in.^2$ (37.91 kPa). The mechanical efficiency of the pump is 0.60 (60%). Assume that the volumetric flow rate of each fluid is 353 ft^3/h (10 m^3/h) at the temperature of 77°F (25°C). Use the computer program "Pumps".

2. Plot the characteristic curves using computed data.

Table 4.1 Data Sheet for Pump Performance

Date: ______________________ Group: ______________________

Type of pump and pump specifications: ______________________

Motor nameplate data:

Fluid used: H_20

Fluid density (ρ): 62.4 lb_m/ft^3 or 1000 kg/m^3

Height of inlet (Z_a):

Height of outlet (Z_b):

Diameter of tube (D):

P_a: 0 psi or kPa

K_h:

Volts:

Time (t):

Amps (@ 3200 RPM):

RPM	Valve open	Power by Watt-Meter, kW	Flowmeter Readings, ft^3 or m^3			Pressure, psi or kPa
			Final	Initial	Actual	
1500	Full					
	1/2					
	1/3					
2000	Full					
	1/2					
	1/3					
3000	Full					
	1/2					
	1/3					
3200	Full					

Table 4.2 Pump Performance Data and Calculations

RPM	Watts	Pressure, lb_f/ft^2 or kPa	Flowrate, ft^3/min or m^3/s	Velocity, ft/min or m/s	Discharge Head, $lb_f ft/lb_m$ or m	Suction Head, $lb_f ft/lb_m$ or m

Table 4.3 Pump Performance Calculations

RPM	Differential Head	Fluid Power, hp or kW	Brake Power, hp or kW	Efficiency, η

Prelab Questions for Laboratory 4

NAME:____________________________ **DATE:**__________________

ANSWER PRELAB QUESTIONS ON THIS SHEET

Laboratory 5

PIPES, FITTINGS, AND PRESSURE DROP MEASUREMENTS

SUMMARY

This lab is intended to acquaint the student with the basic principles and equations of fluid flow including the Reynolds number, laminar and turbulent flow regimes, pressure losses associated with selected valves and fittings, and the determination of overall pumping requirements. Additionally, the student will be introduced to the principles of fluid statics through the use of U-tube manometers.

5.1. BACKGROUND

The pumping of fluid foods through cylindrical pipes is a common means of transporting these products through a food processing facility. Valves and fittings contained along the flow path to facilitate flow regulation, flow diversion, and disassembly for cleaning increase the power required to transport the fluid through the system. To determine the size of a pump for such a system, the pressure drop across each fitting as well as that for the pipe itself and that due to changes in elevation are required. Generally, piping systems are designed to minimize the power required to transport the product. However, certain processing constraints may dictate otherwise. These constraints may include throughput requirements and the intentional creation of turbulence to facilitate energy transfer in heat exchangers or thorough mixing. Optimum flow rates may also be determined in part by quality considerations where high flow rates with large pressure drops may cause physical damage or excessive heat buildup in the product being transported.

5.1.1. Types of Flow

Steady Flow. A system is in a steady state when there is a flow of matter or energy and yet its properties at a given location are invariant with time. Because of this constancy of local conditions, there is no accumulation or depletion of mass or energy within the system and all material and energy balances are of the type

$$\text{Input} = \text{output}$$

When the fluid velocity at each location is constant, the velocity field is invariant with time, and the flow is steady.

One-Dimensional Flow. Velocity is a vector, and in general the velocity at a point has three components (in a three-dimensional system), one for each space coordinate. In many simple situations all velocity vectors in the field are parallel or practically so, and only one velocity component is required. This situation, which obviously is much simpler than the general vector field, is called one-dimensional flow; examples are flows through closed conduits and past plates parallel to the flow. The following discussion is based on the assumptions of steady one-dimensional flow.

Compressible and Incompressible Flow. A fluid possesses a definite density at a given temperature and pressure. If the density is slightly affected by moderate changes in temperature and pressure, the fluid is said to be *incompressible*, and if the density is very sensitive to changes in these variables, the fluid is said to be *compressible*.

Laminar Flow. At low velocities, fluids tend to flow without lateral mixing, and adjacent layers slide past one another like playing cards. There are neither cross-currents nor eddies. This regime is called laminar flow.

Turbulent Flow. Fluids do not flow in laminar motion at higher velocities but move randomly in the form of cross currents and eddies. Such type of motion is turbulent flow.

Reynolds Number. The Reynolds number for the Newtonian fluids is:

$$N_{Re} = \frac{D.\bar{V}.\rho}{\mu} = \frac{D.\bar{V}}{\nu} \tag{5.1}$$

where D = diameter of the tube (m), V = average fluid velocity (m/s), ρ = fluid density (kg/m^3), μ = fluid viscosity (Pa·s), and ν = kinematic viscosity of the fluid, (m^2/s), μ/ρ.

The Reynolds number for non-Newtonian fluids is generally calculated by

$$N_{Re} = 2^{3-n}\left(\frac{n}{3n+1}\right)^n\left(\frac{\rho D^n\,(\bar{V})^{2-n}}{m}\right) \tag{5.2}$$

Where m is the consistency coefficient and n is the flow behavior index for power-law fluids. Equation 5.2 will convert to equation 5.1 for $n = 1$ and $m = \mu$. This is also known as Generalized Reynolds number.

Laminar flow always occurs at N_{Re} below 2100, and the transition from laminar to turbulent flow encountered over a wide range of N_{Re} depending on the tube entrance, and the fluid properties and type. Under normal conditions of flow, a N_{Re} above 4000 provides turbulent flow; and between 2100 and 4000 a transition region occurs. In the transition region, the type of flow may be laminar or turbulent, depending on the tube entrance type and the distance from the entrance. For non-Newtonian fluids, the critical Reynolds number at which transition to turbulent flow starts is given by:

$$N_{Re} = 2100\,\frac{(4n+2)\;(5n+3)}{3\,(3n+1)^2} \tag{5.3}$$

5.1.2. Equation of Continuity (Mass Balance)

In a steady-flow system, the rate of mass entering the "control

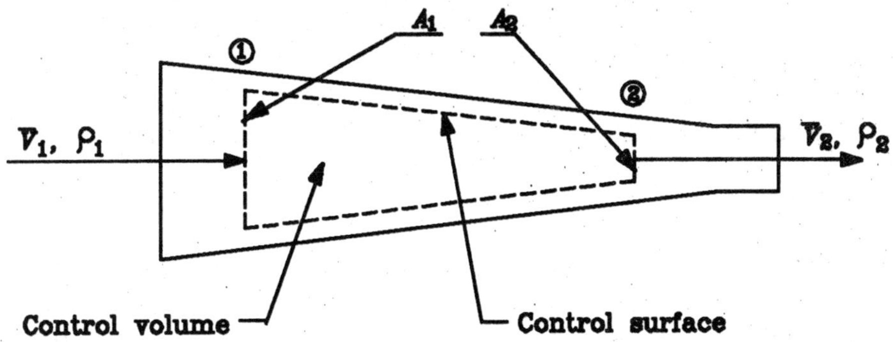

Fig. 5.1 **A typical control volume.**

volume" equals that leaving, as mass can be neither accumulated nor depleted. For an incompressible fluid at steady state, the equation of continuity becomes (Fig. 5.1):

$$\dot{m} = mass\ flow\ rate = \rho_1 \overline{V_1} A_1 = \rho_2 \overline{V_2} A_2 \tag{5.4}$$

5.1.3. Mechanical Energy Balance (Bernoulli's Equation)

Pumping power requirements can be determined by applying a mechanical energy balance between any two points in a system. The mechanical energy balance or Bernoulli's equation for an incompressible fluid at steady state is as follows (for its derivation see any fluid mechanics book):

$$\frac{\alpha_2}{2}\overline{V_2^2} + g.Z_2 = \frac{\alpha_1}{2}\overline{V_1^2} + g.Z_1 - \frac{p_2 - p_1}{\rho} + \delta W_s - 1_v \tag{5.5}$$

where

V	=	fluid velocity
g	=	acceleration due to gravity
Z	=	elevation from a reference
p	=	pressure
ρ	=	fluid density
δW_s	=	shaft work
1_v	=	viscous or frictional losses per unit mass
α	=	1.0 for turbulent flow and 2.0 for laminar flow in circular conduit

Subscript 1 is for location 1 and subscript 2 is for location 2 in equation (5.5).

5.1.4. Friction Losses (Viscous Losses)

Pipeline Losses. The friction loss term in equation 5.5 is a summation of several losses. The first one is the loss due to friction in the tube itself. The losses in a piping network can be broken down into losses in the various straight pipe branches, and then computed with equation 5.5. Consider a horizontal pipe ($Z_2 = Z_1$) of a constant cross-section. If the fluid is incompressible, then from the continuity equation, $V_2 = V_1$. We assume that no shaft work is done in the pipe, so the Bernoulli equation 5.5 simplifies to

$$\frac{p_1 - p_2}{\rho} = l_v \tag{5.6}$$

The pressure difference can be written in terms of friction factor f, a second dimensionless group related to the N_{Re}.

$$f = \frac{p_1 - p_2}{2\rho . V^2} \frac{D}{L} \tag{5.7}$$

where L is the pipe length. Thus, the frictional losses can be expressed as

$$l_v = \frac{2V^2 . L . f}{D} \tag{5.8}$$

To determine this loss, first compute the generalized Reynolds number, and then using Fig. 5.2 or 5.3, determine the friction factor. The friction loss can then be computed from equation 5.8.

Empirical equations for the friction factor as a function of N_{Re} are:

For laminar flow: $$f = 16/N_{Re} \tag{5.9}$$

and

For turbulent flow: $$f = 0.079/N_{Re}^{0.25} \tag{5.10}$$

Loss Due to Sudden Contraction or Expansion. Other friction losses in a fluid transport system are due to contractions or expansions in the tubes. These friction losses can be computed from expressions such as the following for the contraction in a tube:

$$l_v = 0.5V_1^2 . K_f \tag{5.11}$$

where

$$K_f = 0.4\ (1.25 - D_2^2/D_1^2), \text{ at } D_2^2/D_1^2 \leq 0.75 \tag{5.12}$$

$$K_f = 0.75\ (1 - D_2^2/D_1^2), \text{ at } D_2^2/D_1^2 > 0.75 \tag{5.13}$$

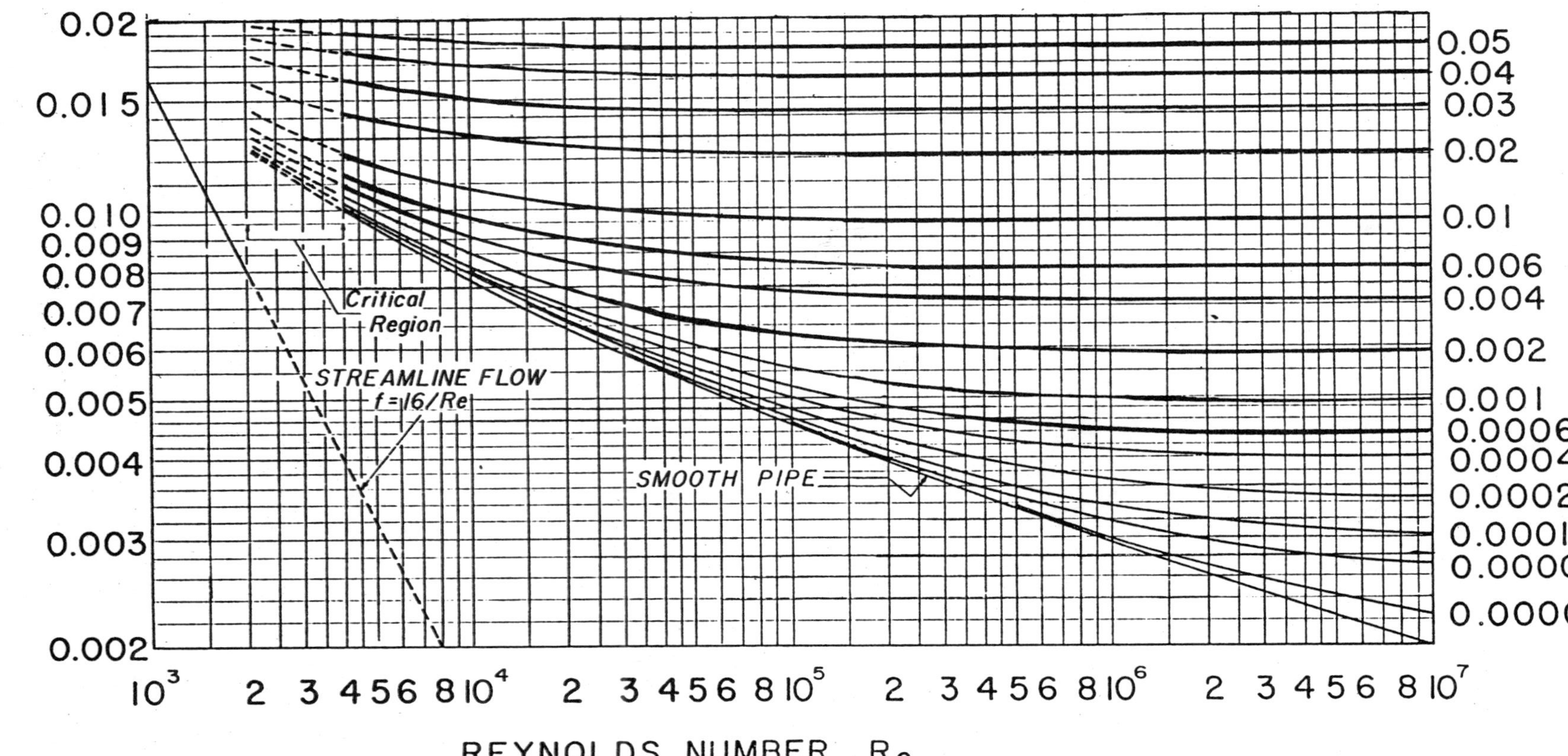

Fig. 5.2 Friction factor as a function of the Reynolds number for incompressible Newtonian fluids (based on Moody, 1944). Equivalent roughness for new pipes (ε in m) are: cast iron, 2.59E-4; drawn tubing, 1.52E-6; galvanized iron, 1.52E-4; steel or wrought iron, 4.57E-5.

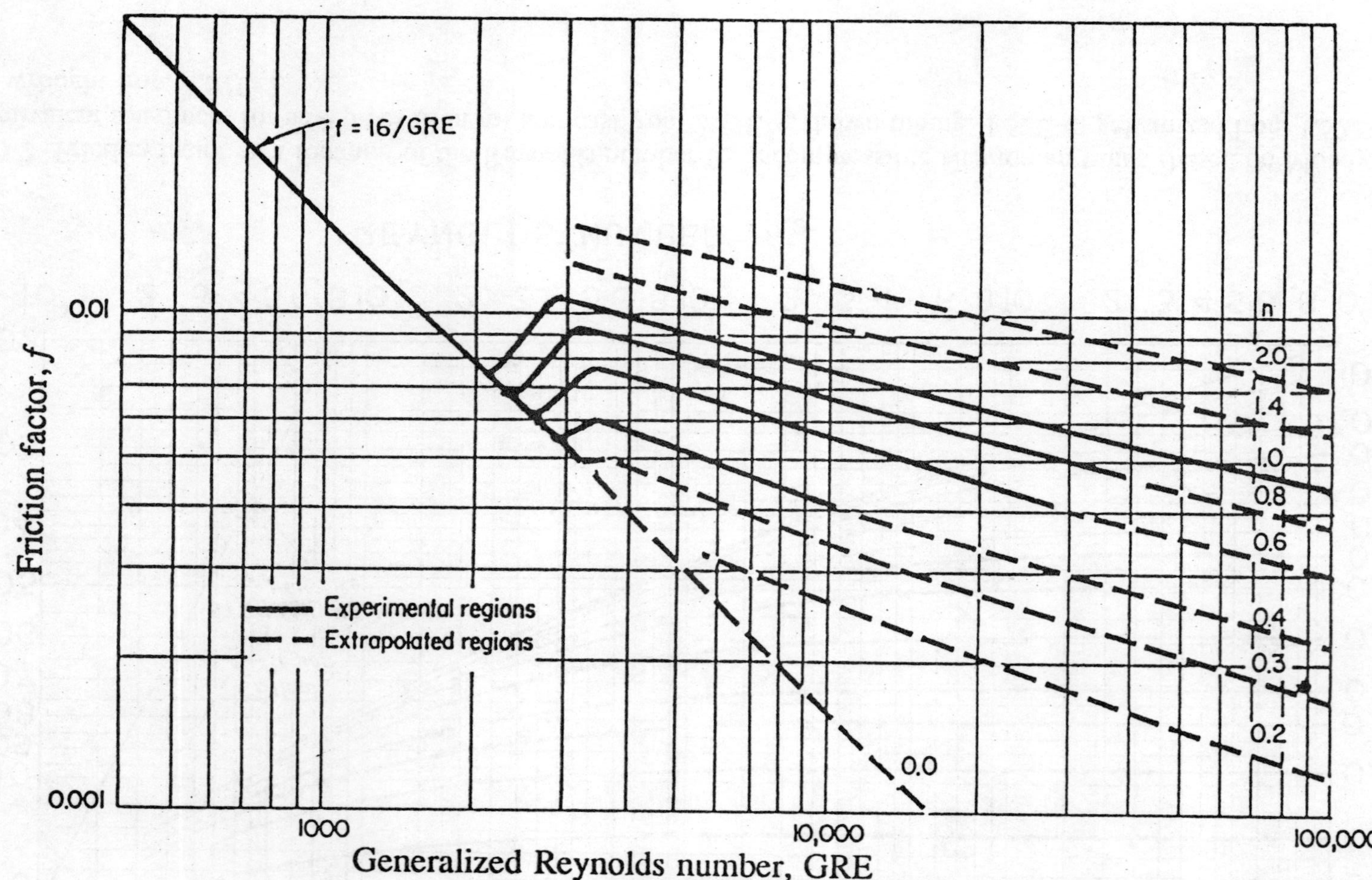

Fig. 5.3 Friction factor vs. generalized Reynolds number chart for non-Newtonian fluid flow in tubes (From Dodge, D.W. and Metzner, A.B. 1959, AIChE J. 5(2):189-204, reprinted with permission of Amer. Inst. Chemical Engineers, New York.)

For an expansion or increase in the tube diameter, the frictional losses can be computed from the following equation:

$$l_v = 0.5V_1^2\,[1 - (A_1/A_2)]^2 \qquad (5.14)$$

where parameters with subscript one are at locations upstream from the expansion or contraction in tube diameter. A is the cross- sectional area of the tube.

Losses Due to Valves and Fittings. The frictional losses in fittings, such as valves, elbows, or tees, can be computed in one of two ways. The first procedure is to utilize experimental data obtained on various types of fittings expressed as equivalent lengths of the tube being utilized. The second approach is to utilize an expression of the following form:

$$l_v = 0.5V^2.K_f \qquad (5.15)$$

in which the constant (K_f) is determined experimentally and would be available as in Table 5.1. It must be emphasized that in both approaches, that of using equivalent length and that of using constant factors in equation 5.15, the available information is for Newtonian fluids. Most available information indicates that the error involved is small.

If we wish to compute the complete losses term for the Bernoulli equation in a pipeline network, we can now simply write

$$l_v = \sum_{\substack{\text{Length}\\ \text{of pipe}}} \frac{2\overline{V_i^2}\,L_i f_i}{D_i} + \sum_{\text{Fittings}} \frac{1}{2}\overline{V_i^2}\,K_{fi} \qquad (5.16)$$

Subscript i refers to the lengths of pipe and fittings that are being summed over. If there is a change in cross-section at the fitting, then V_i refers to the downstream velocity.

5.1.5. Pressure Measurement

The pressure is the force per unit area and can be calculated by $p = \rho.g.h$, where p is the pressure in Pa, ρ is the fluid density in kg/m^3, g is the acceleration due to gravity (9.81 m/s^2) and h is the height of the liquid column in meters. Figure 5.4 shows various pressure measuring configurations: (1) absolute pressure--the reference for this pressure is full vacuum, (2) gauge pressure--the reference is atmospheric pressure, and (3) differential pressure--the reference pressure is another pressure. Thus, all pressure measuring devices operate by determining a change of differential pressure between the measurand and a reference.

Table 5.1 Additional Frictional Loss for Turbulent Flow through Fittings and Valves

Type of Fitting or Valve	Additional Friction Loss, Equivalent No. of Velocity Heads, K_f
45° ell, standard	0.35
45° ell, long radius	0.2
90° ell, standard	0.75
Long radius	0.45
Square or miter	1.3
180° bend, close return	1.5
Tee, standard, along run, branched blanked off	0.4
Used as ell, entering run	1.0
Used as ell, entering branch	1.0
Branching flow	1.0
Coupling	0.04
Union	0.04
Gate valve, open	0.17
3/4 open	0.9
1/2 open	4.5
1/4 open	24.0
Diaphragm valve, open	2.3
3/4 open	2.6
1/2 open	4.3
1/4 open	21.0
Globe valve, bevel seat, open	6.0
1/2 open	9.5
Composition seat, open	6.0
1/2 open	8.5
Plug disk, open	9.0
3/4 open	13.0
1/2 open	36.0
1/4 open	112.0
Angle valve, open	2.0
Y or blowoff valve, open	3.0
Plug cock θ 5°	0.05
10°	0.29
20°	1.56
40°	17.3
60°	206.0

(continued)

Table 5.1 (continued)

Type of Fitting or Valve		Additional Friction Loss, Equivalent No. of Velocity Heads, K_f
Butterfly valve:	5°	0.24
	10°	0.52
	20°	1.54
	40°	10.8
	60°	118.0
Check valve, swing		2.0
Disk		10.0
Ball		70.0
Foot valve		15.0
Water meter, disk		7.0
Piston		15.0
Rotary (star-shaped disk)		10.0
Turbine wheel		6.0

From D.F. Boucher and G.E. Alves, *Fluid and Particle Mechanics* in *Chemical Engineering Handbook* (R.H. Perry and C.H. Chilton, eds.), 1973, McGraw-Hill, Inc. Reproduced with permission of McGraw-Hill, Inc.

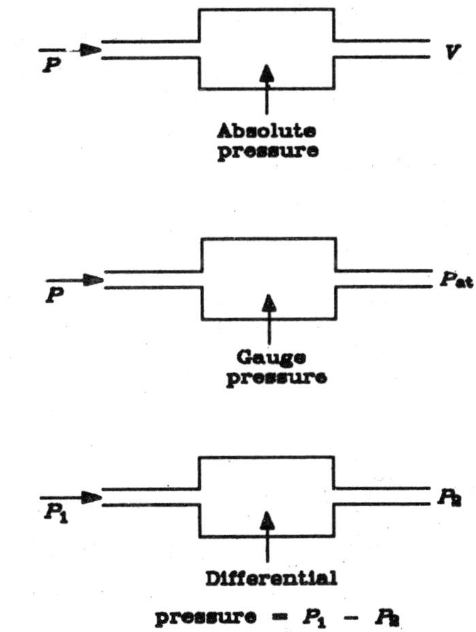

5.4 Pressure measuring configurations. P = input pressure, V = vacuum, P_{at} = atmospheric pressure.

The most common direct pressure transducer is a manometer, the simplest form of which is a U-tube, containing a fluid, which is displaced by the application of different pressures to its ports. For accurate indications a number of corrections and precautions must be taken: (1) the variation of fluid density with temperature, (2) expansion or contraction of attached indicating scale, (3) thermal expansion or contraction of fluid, (4) evaporation of fluid, (5) altitude, (6) fluid contamination, (7) nonverticality of the tubes and scales, and (8) reading difficulty owing to fluid meniscus.

Pressure gauges generally contain a Bourdon tube, which is an elastic pressure element (Fig. 5.5). This has an elliptical form of cross-section and is bent into various configurations: curved-C, spiral, helical, and twisted. The application of a fluid pressure to the open end causes the tube to straighten, thus providing a mechanical displacement. By means of links and gearing, a deflection of less than 3 mm is transformed to a pointer and scale system. The tubes are made of brass, bronze, steel, copper, etc.

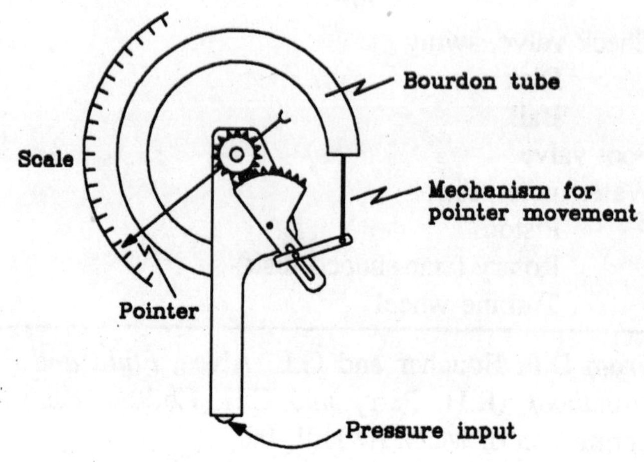

Fig. 5.5 Bourdon tube pressure gauge.

Most pressure transducers require the transduction of pressure information into a physical displacement and means for converting such displacement into a proportional electric signal (Fig. 5.6). The majority of the types depend on the elastic deformation of a diaphragm, a bellow, a Bourdon tube, a spiral, or some combination of these units. The deformation or movement of these elements is transformed to an electrical output signal by one or more of strain gauges, capacitive, piezoelectric, linear variable displacement transducer (LVDT), or resistance-type transducers.

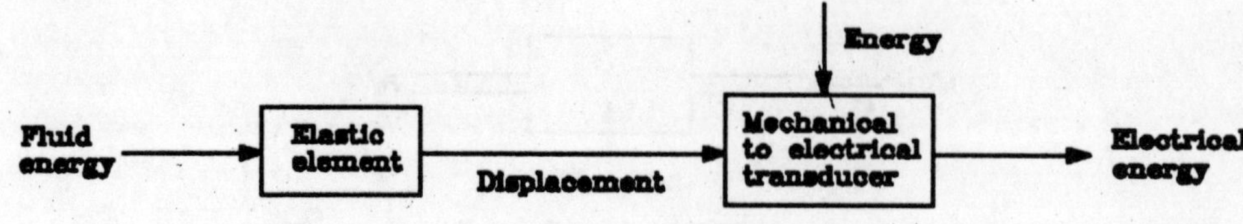

Fig. 5.6 Basic principle of a pressure transducer.

5.2. OBJECTIVES

1. To be familiar with some pipes and fittings that are frequently used in food processing facilities.

2. To obtain experimentally the friction factor vs. Reynolds number relationship for the flow of a Newtonian fluid in a smooth pipe and compare with theoretical results.

3. To experimentally determine the frictional coefficient K_f for two types of valves and several common fittings.

4. To design a piping/ pumping system using a microcomputer which minimizes power requirements while satisfying certain processing constraints.

5.3. IDENTIFICATION OF PIPES AND FITTINGS

You will be measuring the pressure drop across several valves and fittings using a system built specifically for this lab. The flow will be measured with a U-tube manometer and a rotameter. Additionally you will need a milk can, a weighing balance (0 to 50 kg), and a stopwatch.

Prelab Questions

Q1. Would you generally expect air to be compressible or incompressible? What about water?

Q2. If you wanted to increase heat transfer by mixing the fluid, would you want the fluid flowing in the laminar or turbulent regime?

Q3. Would you expect a plot of Δp across a fitting vs. the fluid velocity to be linear?

5.4. PROCEDURE

5.4.1. Pipeline Losses

1. At five different flow rates in the range of 10 to 30 kg water/min, obtain the pressure losses in a gate valve, a compression valve, a sudden contraction, a 90° elbow, a sweeping elbow, and a straight pipe.

2. Determine the exact flow rate in L/min using a weighing balance or scale and a stopwatch. Compare the results with a rotameter.

5.4.2. Design of a Pumping System on a Microcomputer

You are to design a system to heat whole milk prior to the manufacturing of yogurt. The system

must be designed taking into account the following processing constraints. **You are to specify** the route and size of piping and placement of valves and pump. Use the computer program "Pipes and Fittings".
System Constraints

1. Mass flow rate: 2.5 kg/s
2. Fluid density: 1030 kg/m^3 at 20°C (assume constant)
3. Fluid viscosity: 2.62 x 10^{-2} Pa.s at 10°C
   0.69 x 10^{-2} Pa.s at 85°C
4. Holding time: 15 s at 85°C
5. Design for laminar flow except for the holding tubes where the pipe diameter must be 5.08 cm (2 in.) inner diameter
6. Must be able to remove pump, heat exchanger, and holding tube without draining entire system
7. Location of fixed equipment given in the following diagram (Fig. 5.7)
8. Pressure drop in the heat exchanger:
   (a) Heat section = 55 kPa
   (b) Regeneration = 103 kPa
   (c) Cooling section = 69 kPa

5.4.3. Identification of Pipes and Fittings

Identify the following parts and place the appropriate number next to the name of each part.

1. Iron pipe fittings
   ____cross over
   ____cap
   ____reducing elbow
   ____reducer
   ____reducing tee
   ____90° elbow
   ____bushing
   ____plug
   ____reducing cross
   ____street elbow
   ____tee
   ____union
   ____45° elbow
   ____side outlet elbow
   ____lock nut
   ____return bend
   ____elbow union
   ____service tee
   ____cross

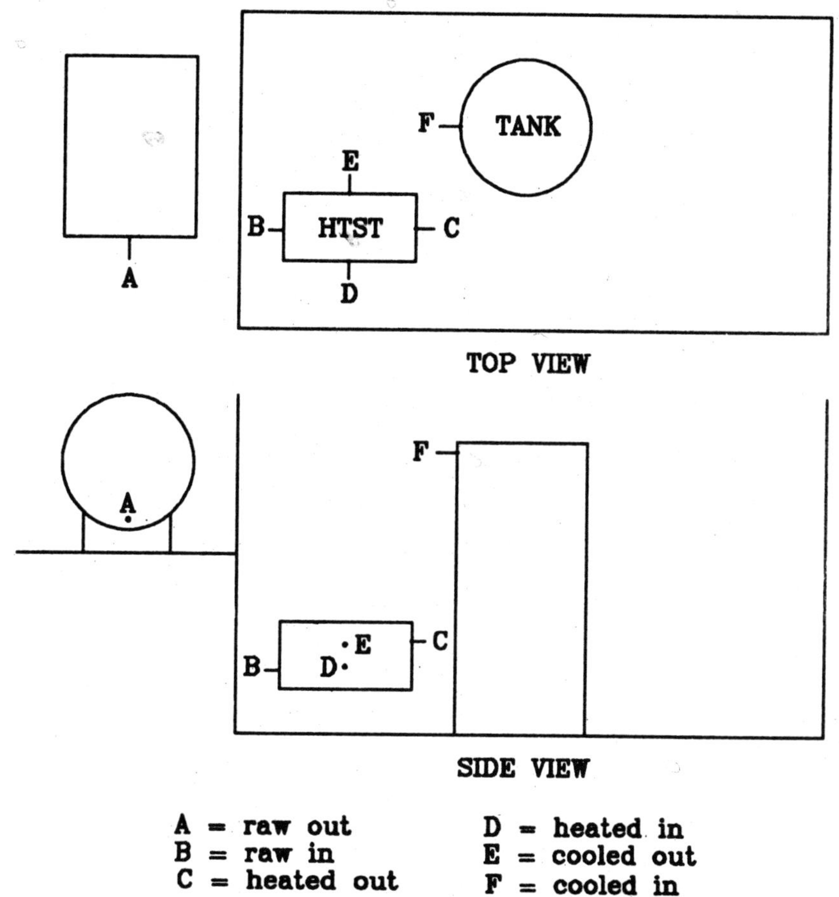

Fig. 5.7 Location of various equipment.

____nipple
____coupling
____strainer
____weld-type fittings
____four-way tee
____floor flange

Iron pipe valves (note assembly of gate, globe, and needle valves)
____swing check valve
____spring-loaded check valve
____butterfly valve
____diaphragm valve

____gate valve
____globe valve
____needle valve

3. Stainless steel pipes and fittings
____clamp
____recessed ferrule
____beveled ferrule
____sanitary pipe
____three-way valve
____two-way valve
____rolled/expanded fitting
____soldered fitting

5.5. REPORT

1. Draw a schematic diagram giving placement of pumps, valves, and pipes for two different system designs.

2. Report the total energy loss and pump power requirements as calculated using a microcomputer.

3. Other than energy requirements what other concerns may affect your design decisions?

5.6. RESULTS AND DISCUSSION

1. Include sheets containing the identification of the selected pipes and fittings.

2. Include a piping diagram of the experimental system.

3. In a table, give the friction factor calculated from your experimental data, the expected friction factor based on experimental data, and the expected friction factor based on the appropriate analytical equation. Calculate the percent error.

4. Plot pressure drop (kPa) vs. V^2 for each of the fittings (if done neatly this may be on one piece of graph paper).

5. From the slope of the plots in part 4, calculate the frictional coefficient, K_f, for each fitting (slope = $K_f/2$).

6. Submit at least two different pipeline/fitting designs (drawings) for the simulated process and identify each with its respective power requirements.

7. Answer the following questions:

   (a) How does the f vs. N_{Re} relationship compare to what is expected? Have you obtained data for both laminar and turbulent flow?
   (b) How do the frictional coefficients obtained experimentally compare with those listed in the reference Table 5.1? What might be several sources of discrepancies?

5.7. POTENTIAL EXPERIMENTAL PROBLEMS

1. Make sure the system is level. Pressures developed by differences in elevation may affect your results.

2. Bleed all the air out of the manometer tubing. The compressibility of gases will cause an error in your pressure equipment.

3. When measuring the pressure drop across the valves, never close both valves at once. This would cause flow to be diverted through the manometer and wash the mercury (which is toxic) down the drain.

4. Extremely low flow rates result in much smaller pressure drops. At such flow rates measurement error becomes a larger problem, flow rates will have to be substantial particularly when measuring the pressure drop in the straight tubing.

Prelab Questions for Laboratory 5

NAME:______________________________________ DATE:________________________

ANSWER PRELAB QUESTIONS ON THIS SHEET

Laboratory 6

MEASUREMENT OF FLOW

SUMMARY

The flow-measuring apparatus is to accustom students to typical methods of measuring the discharge of an essentially incompressible fluid while at the same time giving applications of the steady flow energy equation (Bernoulli's equation). In addition to weigh tank evaluation, the discharge is determined using a Venturi meter, an orifice plate meter and a rotameter. Head losses associated with each meter are determined and compared.

6.1. BACKGROUND

Water enters the apparatus through the lower left-hand end in Fig. 6.1. It flows first through the Venturi meter, then through the orifice meter and so through the rotameter. Connections to the manometer board and relevant dimensions are shown in Fig. 6.1. On leaving the rotameter water flows via a control valve to the weigh tank.

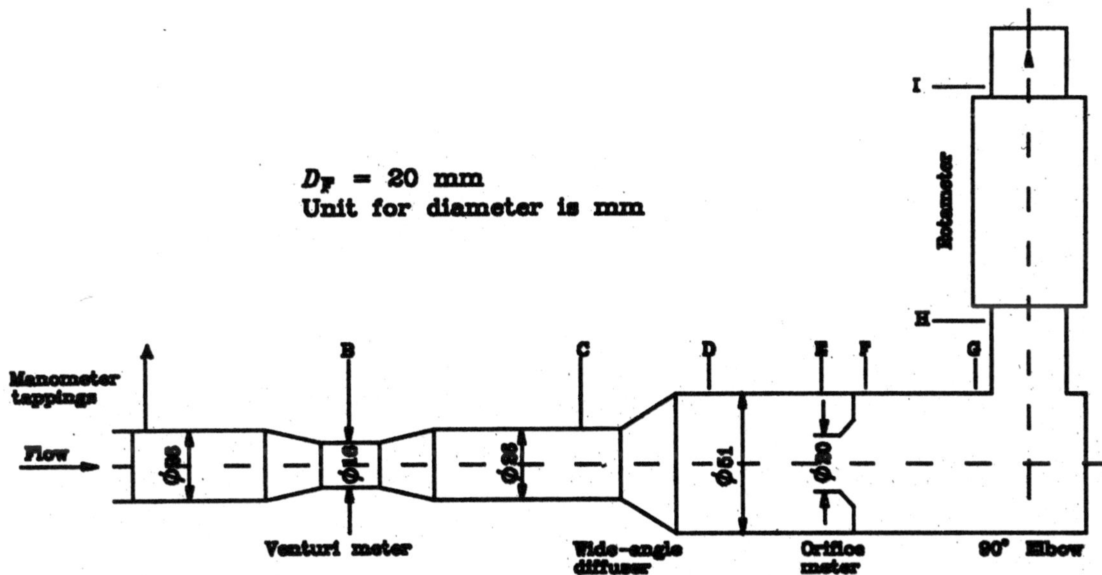

Fig. 6.1 Explanatory diagram of flow-measuring apparatus.

6.1.1. Description of Apparatus

The flow-measuring apparatus is shown in Fig. 6.1. Water enters the equipment through a perspex Venturi meter, which consists of a long gradually converging section, followed by a throat, then by a long gradually diverging section. Pressure measurements in the stream are made at the entry to the meter (*A*), at the throat (*B*), and at the exit (*C*). After a change in cross-section through a rapidly diverging section and another pressure-measuring station (*D*), the flow continues down a settling length and through an orifice plate meter. This meter, manufactured from a brass plate with a hole of reduced diameter through which the fluid flows, is mounted between two pressure-tapped perspex flanges (*E*) and (*F*), (Fig. 6.1) (Tec-Quipment Ltd., Nottingham, England).

Following a further settling length and a right-angled bend the flow enters the rotameter. This consists of a transparent tapered tube in which a float takes up an equilibrium position. The position of this float, assessed from the scale on the wall of the rotameter, is a measure of the flow rate. The pressure drop across the rotameter is given from the manometer levels (*H*) and (*I*).

6.1.2. Measurement of Discharge

The Venturi meter, the orifice plate meter, and the rotameter are all dependent on the modified Bernoulli's equation for their principle of operation.

Bernoulli's equation:

$$\frac{p_1}{\rho} + \frac{V_1^2}{2} + g.h_1 = \frac{p_2}{\rho} + \frac{V_2^2}{2} + g.h_2 + \Delta H_{12} \qquad (6.1)$$

where

p/ρ	=	fluid's hydrostatic head
$V^2/2$	=	fluid's kinetic head
$g.h$	=	fluid's potential head
$p/\rho + V^2/2 + g.h$	=	fluid's total head
ρ	=	density
g	=	9.81 m.s^{-2}

Thus ΔH_{12} is the loss in total head between sections 1 and 2. This term arises because the change in internal energy has not been included in the derivation of Bernoulli's equation and also because the kinetic energy correction factor α has been left out.

Venturi Meter. The modified Bernoulli equation is combined with the continuity equation along with the assumption that ΔH_{12} is accounted for in the discharge coefficient C_v. Therefore, the discharge is calculated using

$$Q = C_v \cdot A_B \sqrt{\frac{2\,(p_A - p_B)}{\rho\,[1 - (D_B/D_A)^4]}} \quad \textit{SI units} \qquad (6.2)$$

where

Q	=	flow rate, m^3/s
C_v	=	discharge coefficient of the Venturi meter
A_B	=	cross sectional area of the throat of the Venturi, m^2
p	=	pressure, Pa
ρ	=	fluid density, kg/m^3
D	=	diameter, m
A,B	=	at location A and B

Orifice Meter. The discharge is given by equation 6.2, C_v for the orifice meter is 0.601, change A to E and B to F in this equation; D_E = 51 mm, D_F = 20 mm.

Rotameter. The rotameter was calibrated by the manufacturer and the calibration curve is provided in the laboratory.

6.1.3. Measurement of Head Loss

Venturi Meter. The nondimensionalized head loss is given by

$$\Delta H_{AC} = \frac{(p_A - p_C)/\rho}{V^2_A/2} = \frac{\text{Head loss}}{\text{Inlet kinetic head}} \qquad (6.3)$$

where

p = pressure, Pa
V = velocity, m/s
ρ = fluid density, kg/m^3

Orifice Meter.

$$\Delta H_{DG} = \frac{(p_D - p_G)/\rho}{V^2_D/2} = \frac{\text{Head loss}}{\text{Inlet kinetic head}} \qquad (6.4)$$

Rotameter.

$$\Delta H_{HI} = \frac{(p_H - p_I)/\rho}{V^2_H/2} = \frac{\text{Head loss}}{\text{Inlet kinetic head}} \qquad (6.5)$$

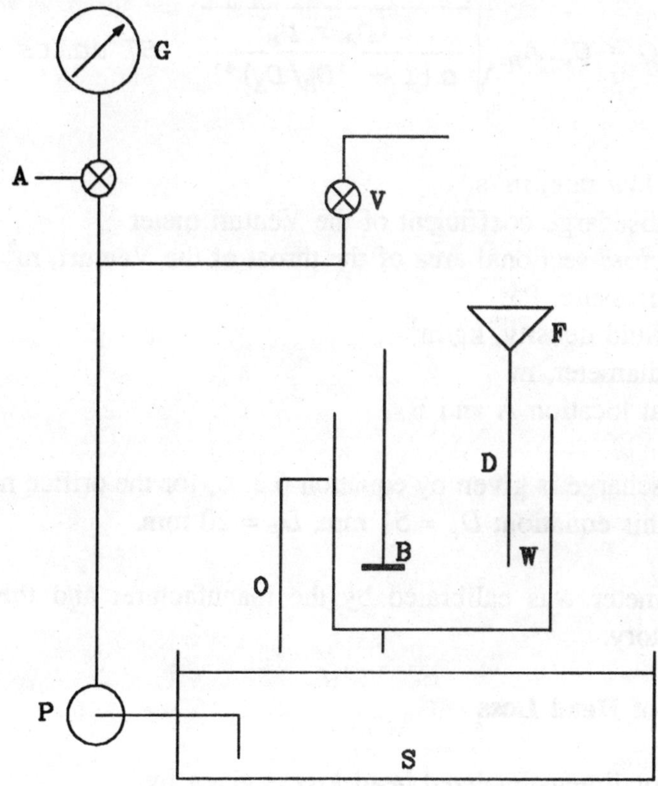

Fig. 6.2 **Diagramatic representation of the single hydraulics bench.**

6.1.4. Description of the Hydraulics Bench

The bench is intended to provide facilities for performing a number of simple experiments in hydraulics. Figure 6.2 shows the arrangement of a single unit in which a small centrifugal pump P draws water from a sump S resting below the bench, and delivers it to a bench supply valve V. The delivery pressure at this valve may be recorded on a Bourdon pressure gauge G, which is provided with a connection A for calibrating the gauge with a dead weight tester (Tec-Quipment Ltd., Nottingham, England). Below the bench there is a weighing tank W into which the discharge for apparatus being tested on the bench may be directed through a short pipe D terminating at flange F just above the bench level. The weighing tank W is supported at one end of a weigh beam, the other end of which carries a weight hanger sufficient to balance approximately the dry weight of the tank. The outlet valve B in the base of the tank may be operated through a mechanism by an external handle. An overflow pipe O is also provided.

There is no permanent connection between the bench top and the supporting framework, thus that the top may be removed easily at any time for inspection of the working parts below. Around the edge of the bench there is a raised lip so that water leaking from apparatus does not spill over the edge, but drains through a waste hole back to the sump. The apparatus under test is placed on the bench and connected by a flexible pipe to a bench supply valve, which normally regulates the rate of flow through the apparatus. Another flexible pipe is led from the exit of the apparatus to the flange above a weighing tank, so that the discharge is returned through the open valve in the base of the tank to the sump. Having set a desired condition in the apparatus, the discharge is normally measured in the following manner. The outlet valve in the base of the weighing tank is closed by operating the handle connected to the valve mechanism. Weights are then applied to the hanger as necessary to tip the weigh beam down to its lower stop at the exposed end of the beam. As water is now steadily collecting in the weighing tank, there comes a time when a weigh beam moves to its upper stop; a stop watch is started at this instant. A further known weight is then added to the hanger, which brings the weigh beam down to its lower stop again. The stopwatch is stopped as the weigh beam moves to its upper stop for a second time, this occurring when the water collected during the timed interval corresponds to the further known weight added. The outlet valve of the weighing tank is then opened to allow it to drain and the weights removed from the hanger in preparation for the next measurement. The weight suitable for collection will vary with the discharge and one or two trials may be necessary at the start of an experiment to determine a suitable weight. An interval of about one minute is normally sufficient to give a satisfactory measurement.

Attention is drawn to the following points that should be observed for the safe and satisfactory operation of the bench.

1. Before starting the pump, ensure that the sump is full and that the bench supply valves are turned off.

2. If a leak develops so that water drips on to the electric motor or starter, stop the pump immediately and isolate it from the electrical supply by withdrawing the plug that supplies it. The connection should not be remade until the leak has been sealed. A small amount of water leaking on to the bench top, however, is of small concern, as it drains back into the sump.

3. When making connections by flexible hose it is usually sufficient to rely on friction between the metal pipe and the hose to maintain the water tightness of the connection. However, where the connection is subjected to the full pressure delivered to the bench supply valve, or if the hose is a loose fit on the metal pipe, it is advisable to secure the connection with a hose clip tightened by a screwdriver or a key made for the purpose.

4. The valves in the base of the measuring tanks should be kept open at all times other than when a discharge measurement is being made. Although each tank has an overflow pipe, this is inadequate to deal with the maximum discharge.

Prelab Questions

Q1. Describe Bernoulli's equation.

Q2. Differentiate between kinetic and potential heads.

Q3. Describe continuity equation.

Q4. What is the concept of "head loss"?

Q5. What is meant by the "discharge coefficient of the Venturi meter"?

Q6. Describe the construction and working of a rotameter.

Q7. Write the equation to calculate flow rate using orifice meter.

6.2. OBJECTIVE

To compare the accuracy and head loss of flow-measuring devices Venturi meter, orifice meter, and rotameter at different flow rates.

6.3. APPARATUS

1. A flow-measuring test bench as discussed or various flow- measuring devices with manometers to measure head loss, pump to supply fluid and a weigh tank.
2. Stopwatch.

6.4. PROCEDURE

The flow-measuring apparatus is connected to the hydraulic bench water supply, and the control valve is adjusted until the rotameter is about at midposition in its calibrated tapered tube. Air is removed from the manometric tubing by flexing it. The pressure within the manometer reservoir is now varied and the flow rate decreased until, with no flow, the manometric height in all tubes is about 280 mm. The apparatus can be leveled and the level checked by comparing the manometric heights when no water flows. A set of experimental readings similar to those in the accompanying Table 6.1 can now be obtained at different flow rates.

6.5. RESULTS AND DISCUSSION

1. Calculate the flow rate and head loss in various flow measuring devices on the attached data sheet (Table 6.1).

2. Plot the flow rate measured by the orifice, Venturi meter, and rotameter vs. flow rate measured by weigh tank on one graph paper. Determine the relative accuracy of these devices by taking slope of the plotted lines.

3. Plot dimensionless head loss for various flow measuring devices vs. inlet kinetic head on one graph paper. Use log-log plot or graph paper to get straight lines.

4. Compare various meters and discuss the relative advantages and disadvantages of each.

Table 6.1 Data Sheet for the Measurement of Flow

I. Raw Data

Test No.	Manometric Levels, mm									Rotameter, reading, cm	Weigh tank (mass of water, kg)	Time, s
	A	B	C	D	E	F	G	H	I			
1.												
2.												
3.												
4.												
5.												
6.												
7.												

Diameters: D_A ________ mm, D_B ________ mm, D_E ________ mm, D_F ________ mm

Area: A_B ________ m^2, A_F ________ m^2 Fluid density ________ kg/m^3

C_v: Venturi meter ________ orifice ________

II. Flow Rate

Test No.	Pressure, Pa									Flow Rate, kg/s			
	A	B	C	D	E	F	G	H	I	Venturi meter	Orifice	Rotameter	Weigh tank
1.													
2.													
3.													
4.													
5.													
6.													
7.													

III. Pressure Drop

Test	Inlet Kinetic Head, J/kg			Nondimensional Head Loss		
No.	Venturi meter	Orifice	Rotameter	Venturi meter	Orifice	Rotameter
1.						
2.						
3.						
4.						
5.						
6.						
7.						

Prelab Questions for Laboratory 6

NAME:______________________________ DATE:______________________

ANSWER PRELAB QUESTIONS ON THIS SHEET

Laboratory 7

HEAT EXCHANGERS

SUMMARY

The transfer of heat from one fluid to another is a common unit operation encountered in food processing. The energy transferred may be in the form of sensible heat or latent heat accompanying a phase change such as condensation or vaporization. The devices used to accomplish this energy transfer are called heat exchangers. The purpose of this lab is to acquaint the student with the fundamentals of heat transfer to fluids and to introduce concepts involved in heat exchanger design and analysis.

7.1. BACKGROUND

7.1.1. Types of Heat Exchangers

Single Pass. In a single-pass heat exchanger, fluid flows through the exchanger only once (Fig. 7.1).

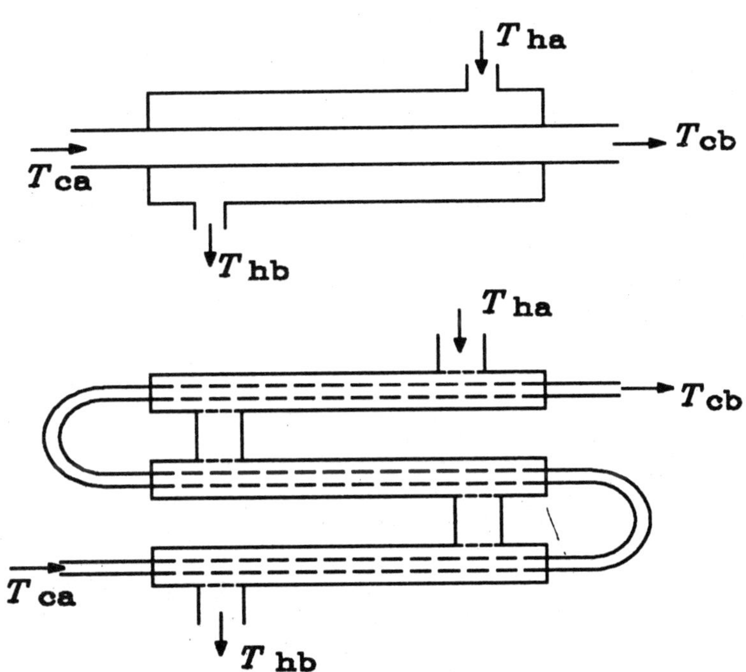

Fig. 7.1 **Double-pipe heat exchangers.**

Countercurrent Flow. Two fluids enter at different ends of the exchanger and pass in opposite directions through the unit. This type of flow is commonly used and is called counterflow or countercurrent flow. The temperature-length curves for this case are shown in Fig. 7.2a. The four terminal temperatures are denoted as follows:

Temperature of entering hot fluid, T_{ha}
Temperature of leaving hot fluid, T_{hb}
Temperature of entering cold fluid, T_{ca}
Temperature of leaving cold fluid, T_{cb}

The approaches are

$$T_{ha} - T_{cb} = \Delta T_2 \text{ and } T_{hb} - T_{ca} = \Delta T_1 \tag{7.1}$$

The warm-fluid and cold-fluid temperature ranges are $T_{ha} - T_{hb}$ and $T_{cb} - T_{ca}$, respectively.

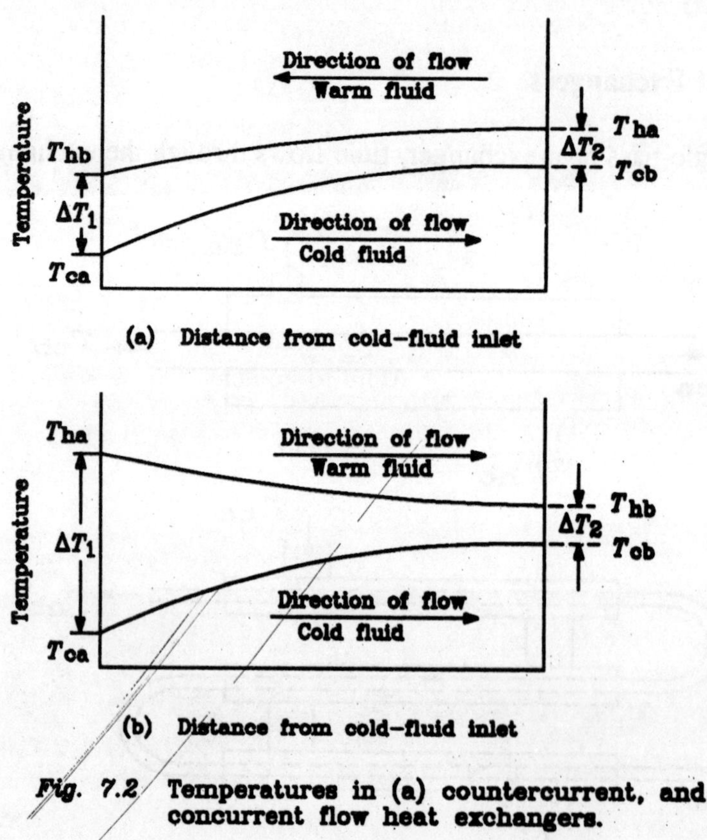

Fig. 7.2 Temperatures in (a) countercurrent, and concurrent flow heat exchangers.

Parallel or Concurrent Flow. If the two fluids enter at the same end of the exchanger and flow in the same direction to the other end, the flow is called parallel. The temperature-length curves for parallel flow are shown in Fig. 7.2b. Again, subscript *a* refers to the entering fluids and subscript *b* to the leaving fluids. The approaches are $\Delta T_1 = T_{ha} - T_{ca}$ and $\Delta T_2 = T_{hb} - T_{cb}$.

Parallel flow is rarely used in a single-pass exchanger because, as inspection of Figs. 7.2a and b will show, it is not possible with this design to bring the exit temperature of one fluid nearly to the entrance temperature of the other, and the heat that can be transferred is less than that possible in countercurrent flow. In multipass exchangers the parallel flow is used in some passes, largely for mechanical reasons, and the capacity and approaches obtainable are thereby affected. Parallel flow is used in special situations where it is necessary to limit the maximum temperature of the cooler fluid or where it is important to change the temperature of at least one fluid rapidly.

Cross-Flow. If the two fluids flow at right angles to one another, the flow is called cross-flow.

Double-Pipe Heat Exchanger. An example of a simple heat-transfer equipment is the double-pipe exchanger shown in Fig. 7.1. It is assembled of standard metal pipes and standardized return bends and return heads, the latter equipped with stuffing boxes. One fluid flows through the inside pipe and the second fluid through the annular space between the outside and inside pipe. Such an exchanger may consist of several passes arranged in a vertical stack as does the one you will use in the lab. Double-pipe exchangers are useful when not more than 9 to 14 m^2 (100 to 150 ft^2) of heat transfer surface is required.

Shell-and-Tube Heat Exchangers. A simple shell-and-tube heat exchanger is shown in Fig. 7.3. It consists essentially of a bundle of parallel tubes--the ends of which are expanded into tube sheets. The tube bundle is inside a cylindrical shell and is provided with two channels, one at each end. Steam or other vapor is introduced into the shell-side space surrounding the tubes, condensate is withdrawn through an outlet, and any noncondensable gas that might enter with the inlet vapor is removed through a vent. The fluid to be heated is pumped through a connection into the channel. It flows through the tubes into the other channel, and is discharged through a connection. The two fluids are physically separated but are in thermal contact with the thin metal tube walls separating them. Heat flows through the tube walls from the condensing vapor to the cooler fluid in the tubes.

Plate Heat Exchangers. For heat transfer between fluids at low or moderate pressures, below about 2 MPa (20 atm) pressure, plate-type exchangers are competitive with shell-and-tube exchangers, especially where corrosion-resistant materials are required. Metal plates, usually with corrugated faces, are supported in a frame; hot fluid passes between alternate pairs of plates, exchanging heat with the cold fluid in the adjacent spaces. The plates are typically 5 mm apart. They can be readily separated for cleaning; additional area may be provided simply by adding more plates. Unlike shell-and-tube exchangers, plate exchangers can be used for multiple duty (i.e., several different fluids can flow separately through different parts of the exchanger).

Scraped Surface Heat Exchangers. Viscous liquids and liquid-solid suspensions are often heated or cooled in scraped surface exchangers. Typically these are double pipe exchangers with a fairly large central tube 10 to 30 cm (4 to 12 in.) in diameter, jacketed with steam or cooling liquid. The inside surface of the central tube is wiped by one or more longitudinal blades mounted on a rotating shaft.

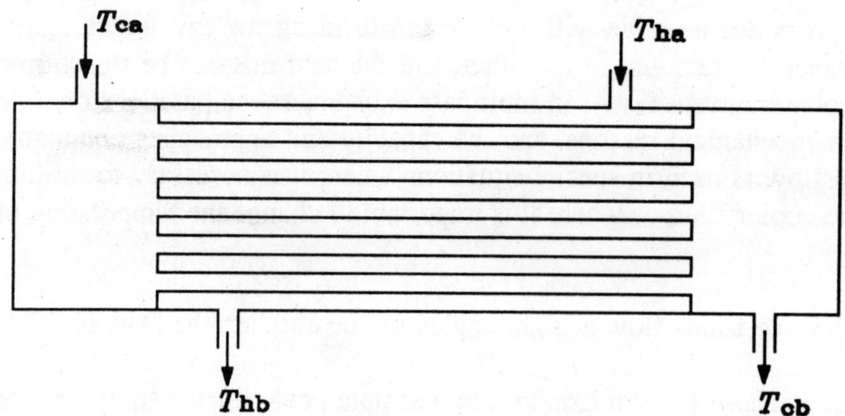

Fig. 7.3 **Shell-and-tube heat exchanger, single pass.**

The viscous liquid is passed at low velocity through the central tube. Portions of this liquid adjacent to the heat transfer surface are essentially stagnant, except when disturbed by the passage of the scraper blade. Heat is transferred to the viscous liquid by unsteady-state conduction. If the time between disturbances is short, as it usually is, the heat penetrates only a small distance into the stagnant liquid, and the process is exactly analogous to unsteady-state heat transfer to a semiinfinite solid.

7.1.2. Heat Flux

Heat transfer calculations are based on the area of the heating surface and are expressed in W/m^2 or $Btu/(h.ft^2)$ of surface through which the heat flows. The rate of heat transfer per unit area is called the heat flux. In many types of heat-transfer equipment the transfer surfaces are constructed from tubes or pipe. Heat fluxes may then be based either on the inside area or the outside area of the tubes. Although the choice is arbitrary, it must be clearly stated because the numerical magnitude of the heat fluxes will not be the same for both.

7.1.3. Overall Heat Transfer Coefficient

It is reasonable to expect the heat flux to be proportional to a driving force. In heat flow, the driving force is $T_h - T_c$, where T_h is the average temperature of the hot fluid and T_c is that of the cold fluid. The quantity $T_h - T_c$ is the overall local temperature difference ΔT. It is clear from equation 7.1 that ΔT can vary considerably from point to point along the tube, and, therefore, since the heat flux is proportional to ΔT, the flux also varies with tube length. It is necessary to start with a differential equation, by focusing attention on a differential area dA through which a differential heat flow dq occurs under the driving force of a local value of ΔT. The local flux is then dq/dA and is related to the local value of ΔT by the equation

$$\frac{dq}{dA} = U\,\Delta T = U(T_h - T_c) \tag{7.2}$$

The quantity U, defined by equation 7.2 as a proportionality factor between dq/dA and ΔT, is called the local overall heat transfer coefficient.

The overall heat transfer coefficient may also be defined as follows:

$$U = \frac{q}{A\,\Delta T} = \frac{\Delta T/\Sigma R}{A\,\Delta T} = \frac{1}{A\,\Sigma R}$$

or, for the case of a tubular heat exchanger,

$$U = \frac{1}{A[1/(A_i h_i) + [\ln(r_o/r_i)]/(2\pi kL) + 1/(A_o h_o)]} \tag{7.3}$$

Here, ΣR represents the sum of all the resistances to heat flow. Equation 7.3 indicates that the overall heat-transfer coefficient U may have a different numerical value, depending on which area it is based upon. If, for instance, U is based on the outside surface area of the pipe, A_o, we have

$$U_o = \frac{1}{A_o[1/(A_i h_i) + [\ln(r_o/r_i)]/(2\pi kL)] + 1/h_o} \tag{7.4}$$

thus it is necessary, when specifying an overall coefficient, to relate it to a specific area. Table 7.1 lists U values for a few typical systems.

7.1.4. Heat Transfer Analysis

The heat transfer (q) between fluids in a single-pass exchanger is calculated by

$$q = U.A.\Delta T_{lm} \tag{7.5}$$

where ΔT_{lm} is known as logarithmic mean temperature difference, and is given by

$$\Delta T_{lm} = \frac{\Delta T_2 - \Delta T_1}{\ln\left(\frac{\Delta T_2}{\Delta T_1}\right)} \tag{7.6}$$

Table 7.1 Approximate Values for Overall Heat Transfer Coefficients

Fluid Combination	U, Btu/(h.ft^2.°F)	W/(m^2.K)
Water to compressed air	10-30	57-170
Water to water, jacket water coolers	150-275	852-1560
Water to brine	100-200	570-1135
Water to gasoline	60-90	340-511
Water to gas oil or distillate	35-60	200-340
Water to organic solvents, alcohol	50-150	284-850
Water to condensing alcohol	45-120	255-680
Water to lubricating oil	20-60	114-340
Water to condensing oil vapors	40-100	227-570
Water to condensing or boiling Freon-12	50-150	284-850
Water to condensing ammonia	150-250	852-1420
Steam to water, instantaneous heater	400-600	2270-3400
storage-tank heater	175-300	994-1700
Steam to oil, heavy fuel	10-30	57-170
light fuel	30-60	170-340
light petroleum distillate	50-200	284-1135
Steam to aqueous solutions	100-600	570-3400
Steam to gases	5-50	28-284
Light organics to light organics	40-75	227-426
Medium organics to medium organics	20-60	114-340
Heavy organics to heavy organics	10-40	57-227
Heavy organics to light organics	10-60	57-340
Crude oil to gas oil	30-55	170-312

From J.R. Welty, C.E. Wicks and R.E. Wilson, *Fundamentals of Momentum, Heat and Mass Transfer*, COPYRIGHT © 1984, John Wiley & Sons, Inc. Reprinted by permission of John Wiley & Sons, Inc.

Prelab Questions

Q1. What is the lowest temperature attainable for the hot fluid in a counter flow heat exchanger?

Q2. Is turbulence easier or harder to achieve in a plate heat exchanger as compared to a double-pipe exchanger?

Q3. By analogy to the log mean temperature difference, how would you calculate a log mean area?

7.2. OBJECTIVES

The objectives of this lab are to:

1. Observe the differences between countercurrent and parallel flow in the performance of a double-pipe heat exchanger.

2. Determine the effect of varying flow rate on the overall heat transfer coefficient and the heat flux in a double-pipe heat exchanger.

3. Determine the effect of varying the consistency coefficient and flow behavior index of nonNewtonian fluids on the convective heat transfer coefficient and the required length of a heat exchanger using a microcomputer.

7.3. APPARATUS

1. Double-pipe heat exchanger with two rotameters and digital thermocouple display or temperature recorder or data acquisition system

2. 0 to 50 kg (0 to 100 lb), weighing balance or scale

3. Stopwatch

7.4. PROCEDURE

7.4.1. Double-Pipe Heat Exchanger

1. Have both hot and cold water flow in the same direction (parallel) at a flow rate of approximately 11 L/min (3 gpm).

2. When steady state is reached, read and record temperatures on the data sheet (Table 7.2) provided.

3. Increase the cold flow to about double the former rate, allow the temperatures to equilibrate, and record temperatures on the Table 7.2.

4. Reverse the cold water flow and repeat steps 1 through 3.

7.4.2. Heat Exchanger Design for Non-Newtonian Fluids

Using the computer program "Heat Exchangers", determine the effects of varying consistency coefficient m and flow behavior index n of nonnewtonian fluids on several design characteristics of double-pipe heat exchangers.

1. Influence of flow behavior index, *n*

The following data describes the process:

Product temperature

Initial 10°C

Final 80°C

Heating medium temperature: 121°C

Mass flow rate of the product: 20 kg/s

Product Characteristics

	m, (Pa.s^n)	n
Honey	5.6	1.0
Orange juice concentrate	5.3	0.73
Applesauce	5.6	0.47
Pear puree	5.3	0.38
Apricot puree	5.4	0.29
Thermal conductivity	0.410 W/(m.K)	
Density	1090 kg/m^3	
Specific heat	3750 J/(kg.K)	

Run the computer program. Input process and product specifications when prompted by the microcomputer. After the computer returns a table of heat transfer coefficients and heat exchanger lengths, press shift and print-screen keys to obtain a hard copy of the table. Include this with your lab report.

2. Influence of Consistency Coefficient, *m*

The product temperatures and flow rates as well as density, thermal conductivity and specific heat are the same as in part 1.

	m, (Pa.s^n)	*n*
Tomato concentrate	8.7	0.4
Apricot puree	7.2	0.4
Guava puree	4.2	0.4
Applesauce	0.66	0.4

Again, obtain a table of values for heat transfer coefficients and heat exchanger lengths for different values of *m* and include this with your lab report.

In the foregoing analyses, the following assumptions and constraints apply.

1. Resistance to heat transfer through the tube material and through the film on the heating-medium side are negligible compared to the product side.

2. Fluid properties do not change with temperature (is this a valid assumption?).

3. Initially, the largest tubing size that will allow turbulent flow is calculated, then, using the same pipe diameter, the lowest velocity that permits turbulent flow is calculated for the effect of m.

7.5. RESULTS AND DISCUSSION

7.5.1. Double-Pipe Exchanger

1. Plot both hot and cold flow temperatures vs. length for the four runs on separate graphs.

2. Calculate the amount of heat transferred from the hot water for each run.

3. Calculate the log mean temperature difference, and by analogy the log mean area.

4. Calculate the overall heat transfer coefficients, U, for each run based on the results of part 3.

5. Tabulate the results from part 2 through 4 and discuss.

7.5.2. Heat Exchange in Non-Newtonian Fluids

1. Plot h_c vs. m, h_c vs. n, length vs. m and length vs. n.

2. Briefly discuss the effects of m and n on heat exchanger design.

7.6. POTENTIAL EXPERIMENTAL PROBLEMS

Please be careful as some of the heat exchange surfaces may be hot, especially if you are using steam as the hot fluid. When using the chilled water system, the flow must be returned to form a closed circuit.

Table 7.2 Heat Exchangers Data Sheet

							Parallel Flow								
							Temperature, °C								
		COLD							HOT						
	Flow Rate, kg/min	1	2	3	4	5	6	7	8	9	10	11	12	13	14
1. Run															
Hot															
Cold															
2. Run															
Hot															
Cold															
							Counterflow								
3. Run															
Hot															
Cold															
4 Run															
Hot															
Cold															

Tube dimensions, cm	I.D.	O.D.
Inner		
Shell		
Pipe length:	________	
Distance between thermocouples:	________	

Prelab Questions for Laboratory 7

NAME:____ **DATE:**__________________

ANSWER PRELAB QUESTIONS ON THIS SHEET

Laboratory 8

APPLICATIONS OF PLATE HEAT EXCHANGERS

SUMMARY

By now you have been exposed to the principles of fluid flow (Labs 3 to 6) and basic heat exchange (Lab 7). The goal of this lab is to demonstrate how these principles are brought together in the design of a food processing operation through evaluation of the performance of a plate heat exchanger.

8.1. BACKGROUND

8.1.1. HTST Pasteurization

A number of uses of plate heat exchangers exist in the food industry. The most common is perhaps their use for high- temperature, short-time (HTST) pasteurization of milk and milk- based products. For HTST pasteurization of milk, the Milk Ordinance and Code, Recommended Practices of the United States Public Health Service requires that the product be heated to 161°F (71.7°C) and held at or above that temperature for 15 s. For ice cream mix the requirements are 175°F (79.4°C) for 25 s. Plate heat exchangers used for HTST pasteurization are generally equipped with sections for regenerative heating, heating, and cooling. Typical schematics of HTST Systems are shown in Figs. 8.1 and 8.2.

It is the practice in North America to use a positive displacement pump to feed the heater. This provides positive flow control and a higher pressure in the pasteurized regenerator than in the raw (so that leaks are from safe product to product about to be pasteurized). The European practice is to use a self-regulating rotameter-flow control valve or a constant pressure drop (ΔP) orifice to control flow and a centrifugal pump. This pump may then feed the generator section.

For start-up especially with high regeneration (i.e., energy recovery), it is necessary to provide a bypass around the regenerator to speed apparatus come-up. A leak is provided in the closed bypass valve to avoid having a "dead" space in the bypass line. For high regeneration, a booster pump feeding the regenerator is required for the North American approach, but this will then require a differential pressure switch from raw regenerator to pasteurized regenerator out to assure 6.9 kPa (1 psi) pressure differences. Regeneration values in North America approach 95% at an appreciable cost in added plates and additional pumping energy.

Homogenizers are placed ahead of the holding tube for pasteurizers, after the flash cooler for steam injection final heaters, and after the first of two regenerator cooling sections for indirect sterilization. Homogenizers must have a flow loop from discharge back to their feed to avoid being a flow promotion device if used with a positive pump upstream of the regenerator for a simple system. Holding sections are tubular with a continuous rise in a trombone or circular

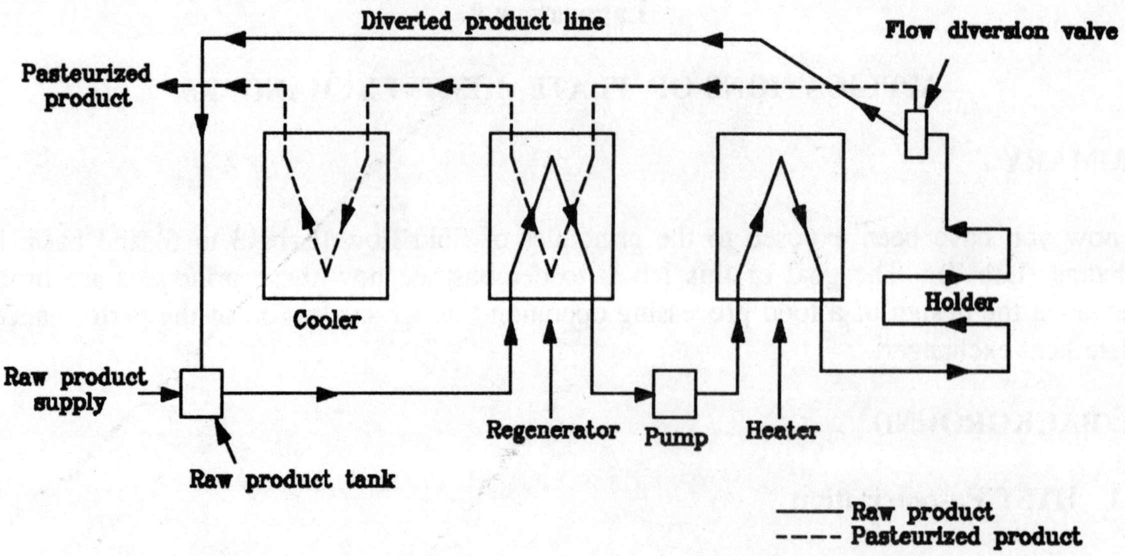

Fig. 8.1 **A simple HTST pasteurizing system. Heating and cooling medium lines and temperature control equipment are not shown.**

array in North America. Heat losses are often minimal, but a holding tube may be enclosed. A system may have two holding tubes in series or as alternative routes to give higher heat treatment desired for some products.

A check thermometer and the controller thermometer are placed just ahead of the holding tube and the flow diversion valve just after it for pasteurizers. For sterilizers, the flow diversion valve may stay there with two flow diversion sensors, one at the holding tube entrance and the other at the holding tube exit, to allow for slow sensor response for the sterilizers. Alternatively, they may be placed at the end of the regenerator to allow the holding tube exit sensor (controller) valve with booster relay to divert for under-temperature-high-temperature product.

To avoid foaming, which one encounters in milk at 4°C to 6°C, a second cooler may be used. Its coolant is usually a food-grade glycol. Its controller is important to avoid freezing damage, but parallel flow of glycol at 0.5°C helps reduce this danger. The first cooler (or regenerator) may be split to provide cream separation at a desirable temperature with a separate cream cooler downstream. A vent to atmosphere is required to keep the centrifuge from being a flow promoting device that could change the holding time.

Pasteurizers in North America have internal drain holes in the bottom of the plates to permit product to drain after operation. The bottom of the plates is 30.5 cm (1 ft) above the ballast tank level to assist in this drainage, and the product discharge line is placed 30.5 cm (1 ft) above the highest point in the system then vented to the atmosphere, also to permit drainage.

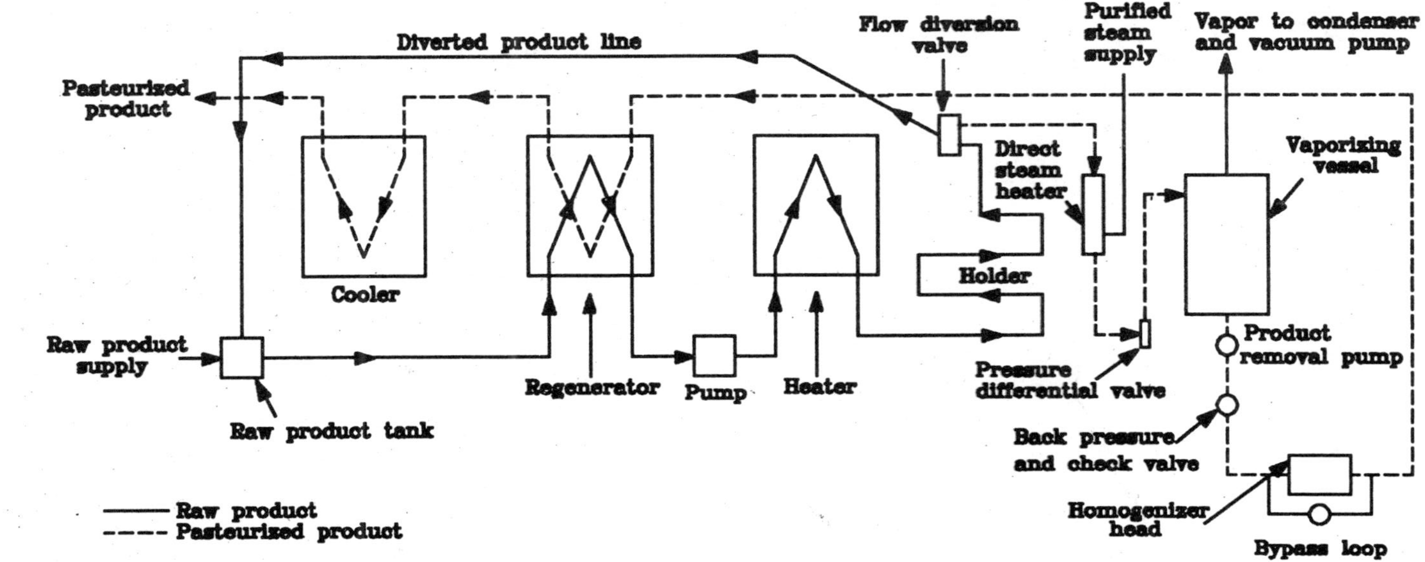

Fig. 8.2 An HTST system incorporating steam–vacuum, flavor–treating equipment. The homogenizer is downstream from the flavor–treating equipment.

Often, a throttle restriction is placed in the discharge line to "swamp" the effect on pressures of changing from one product tank or filter to another.

8.1.2. Thermal Performance

In plate heat exchangers, the closely spaced heat transfer plates have troughs or corrugations that induce turbulence in the liquids flowing as a thin stream between them. Unlike the round closed conduits previously discussed where the transition from laminar to turbulent flow occurs at N_{Re} of 2100 to 4000, the N_{Re} at which flow becomes laminar in a typical plate heat exchanger may be as low as 100 to 400 depending on the type of plate. This makes plate type heat exchangers particularly efficient in handling viscous liquids because the critical N_{Re} is low. For plate heat exchangers, the overall heat transfer coefficient is proportional to $(N_{Re})^m$, since Nu (Nusselt number, $h.L/k$) = $c_1.N_{Re}{}^m.Pr^a$, where c_1, m, and a are constants; h is the heat transfer coefficient, k is the fluid thermal conductivity, L is the characteristic length, and Pr is the Prandtl number. Remembering that $N_{Re} = \bar{V}.D.\rho/\mu$, this implies that U is proportional to $\bar{V}^m$ at constant temperature. Typical average velocities in plate heat exchangers for waterlike fluids in turbulent flow are 0.3 to 0.9 m/s, but true velocities in certain regions will be higher by a factor of up to four due to the effect of corrugations. Heat transfer relationships however, are generally based on either a velocity calculated from the average plate gap or on the flow rate per passage. Often the parameter of interest is the throughput or mass flow rate. Since $\dot{m} = \rho.\bar{V}.A$, and for varying throughput and constant temperature, ρ and A are constant; we can also see that U is proportional to $(\dot{m})^m$.

8.2. Holding Time

If a product is to receive a required pasteurization process, it is necessary to control the temperature history. The simplest approach is to maintain the product at a constant temperature for a prescribed length of time. We know that, for a tube in which all fluid has the same velocity,

$$\text{Holding time } (t) = \text{length } (L)/\text{velocity } (V) \qquad (8.1)$$

Furthermore, from the requirement that mass flow into a tube equals mass flow out from the tube,

$$V = \dot{m}/\rho.A = Q_v/A \qquad (8.2)$$

where

V = mean fluid velocity
$\dot{m}$ = mass flow rate
ρ = density of the fluid
Q_v = volumetric flow rate
A = cross-sectional area of the tube

Hence we would expect that $t = L.A/Q_v$, if all the particles go through a tube at the same rate, but they do not. Hence, the approach outlined in the USPHS handbook assumes that an experimental measurement is required, which is the experimental time required for the fastest detectible microaggregate of salt and matter of molecules to pass from the point of injection to the point of detection.

What can one do if the time is too short to measure? This occurs for ultrahigh temperature and sterilization operations. These times are in fractions of second. If we knew the relationship between the velocity of the fastest and the average particles (and all particles moved in straight lines), we could determine the minimum holding time (t_{MIN}):

$$t_{MIN} = (L.A/Q_v)/(V_{max}/V_{mean}) = L/V_{max} \tag{8.3}$$

For laminar flow, the fastest particles (at the center) travel twice as fast as the average. This is true for Newtonian fluids that flow in circular conduits. When the flow is turbulent the bulk of

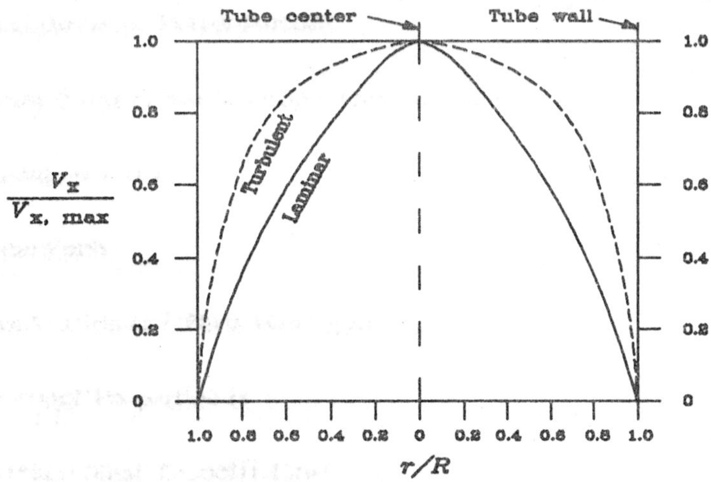

Fig. 8.3 **Qualitative comparision of laminar and turbulent velocity distributions.**

the fluid, outside the boundary layers, is at a nearly constant velocity. The maximum velocity is of the order of only 1.2 times that of the average (Fig. 8.3). This ratio depends on the Reynolds number.

8.3. OBJECTIVES

1. To observe the effect of fluid velocity (mass flow rate) on the overall heat transfer coefficient U in a plate heat exchanger.

2. To determine residence times (holding times) in an HTST pasteurizer using a conductivity based holding-time timer and by calculation from flow rate data.

8.4. APPARATUS

1. A plate heat exchanger with heating, regeneration and cooling sections, and holding tube.

2. A conductivity-based, holding-time timer, consisting of two conductivity probes and a syringe for salt injection, with sanitary pipe tees to allow their installation.

3. Zero to 50 kg (0 to 100 lb) weighing scale.

4. Stopwatch.

8.5. PROCEDURE

1. For three different product mass flow rates, record the inlet and outlet temperatures of the heating medium and product.

2. For the same flow rates as in 1, determine the holding times using the conductivity based timer.

8.6. RESULTS AND DISCUSSION

1. Calculate the amount of heat transferred based on the temperature change of the product.

2. Calculate the log mean temperature difference.

3. Calculate U based on the answers for 1 and 2 above and the heat transfer area.

4. Plot U vs. $\dot{m}$.

5. Plot log (U) vs. log ($\dot{m}$) and determine the value of n in the expression "U proportional to $(\dot{m})^n$."

6. Plot holding time vs. $\dot{m}$ for both the calculated and measured times. Do they give the same values?

Prelab Questions

Q1. What is the purpose of pasteurization? Is the final product sterile? How were the temperatures of 161 and 175°F (71.7°C and 79.4°C) determined? Why should ice cream mix require a higher heat treatment?

Q2. Define regeneration.

Q3. For a holding tube 15.24 m (50 ft) long and 5.08 cm (2 in.) I.D. through which a fluid flows at 2.28 m/s (7.5 ft/s), what is the average residence time of a particle in the tube? What is the residence time of the fastest particle if the fluid is milk?

8.7. POTENTIAL PROBLEMS

We will be using steam as a heating medium. Please be careful, the potential for serious burns exists.

Saturated steam 140°C

Condenser

Condensate 110°C

Vapor 40°C

Warm water

Dilute peach puree, 25°C, 12% solids

Evaporator

Concentrated peach puree 100 kg/h 40°C, 40% solids

Fig. 2.2 Schematic diagram for Problem 2.7.

Prelab Questions for Laboratory 8

NAME:____________________________ **DATE:**____________________

ANSWER PRELAB QUESTIONS ON THIS SHEET

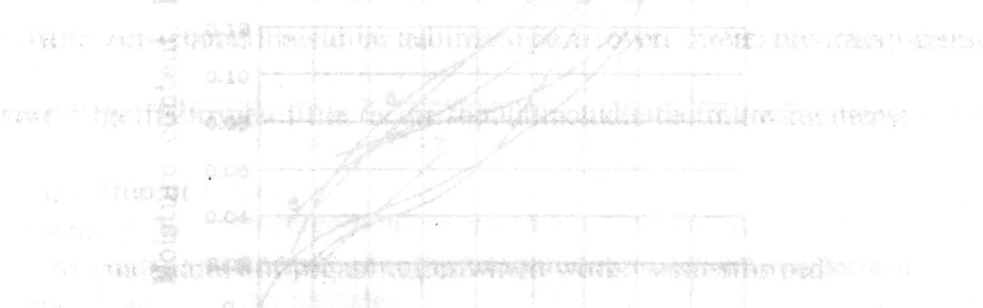

Laboratory 9

THERMAL PROPERTIES OF FOODS

SUMMARY

The thermal properties of food materials such as specific heat, enthalpy, thermal conductivity, and thermal diffusivity are important for efficient design of processes involving heat transfer. Several examples of such processes are sterilization, pasteurization, evaporation, concentration, drying, dehydration, freezing, and refrigeration. The temperature span of interest is generally -50°C to 150°C, which covers two-phase changes in water at atmospheric pressure, of which freezing is associated with dramatic changes in thermal properties. In this lab you will determine the specific heat, thermal conductivity, and thermal diffusivity of selected foods.

9.1. BACKGROUND

9.1.1. Specific Heat

The specific heat of a material is the amount of heat required to increase the temperature of a unit mass by 1°C or °F or K. The specific heat of a substance can be determined using a simple adiabatic calorimeter where the change in temperature of a known amount of material of unknown specific heat is compared to the change in temperature of a known amount of a reference material of known specific heat after the two have been mixed and the temperature allowed to equilibrate. Assuming that the total thermal energy in a vacuum flask, a charge of water (reference material), and a food sample remains constant before and after mixing them,

$$T_iC_fW_f + T_iC_cW_c + T_sC_sW_s = (C_fW_f + C_cW_c + C_sW_s)T_m \qquad (9.1)$$

where

T = temperature
W = mass
C = specific heat
i = initial time
f = flask
c = water charge
s = sample
m = mixture

Upon rearrangement, this yields, for C_s,

$$C_s = (C_fW_f + C_cW_c)\ (T_m - T_i)/(W_s(T_s - T_m)) \qquad (9.2)$$

9.1.2. Thermal Conductivity

Thermal conductivity k arises as the proportionality factor in Fourier's law of heat transfer by conduction, stated as follows (considering ρ, C_p, and k independent of T):

$$\rho.C_p . \frac{\delta T}{\delta t} = k.\nabla^2 T \tag{9.3}$$

For an isotropic and one-dimensional system (where k is not a function of the direction of heat transfer), this reduces to:

$$\rho.C_p.\frac{\delta T}{\delta t} = k.\frac{\delta^2 T}{\delta x^2} \tag{9.4}$$

where

ρ	=	density
C_p	=	specific heat
T	=	temperature
t	=	time
k	=	thermal conductivity
x	=	distance in the x-direction

Dividing each side of equation 9.4 by $\rho.C_p$ we obtain the "heat transfer equation" in one dimension:

$$\frac{\delta T}{\delta t} = \frac{k}{\rho.C_p}.\left(\frac{\delta^2 T}{\delta x^2}\right) \tag{9.5}$$

The value $k/(\rho.C_p)$ is thermal diffusivity and is abbreviated by α. The thermal diffusivity of a material is the ratio of its heat conducting capacity to its heat capacity. The form of the heat transfer equation in cylindrical coordinates is:

$$\frac{\delta T}{\delta t} = \alpha.\left(\frac{\delta^2 T}{\delta r^2} + \frac{1}{r}.\frac{\delta T}{\delta r} + \frac{1}{r^2}\frac{\delta^2 T}{\delta \theta^2} + \frac{\delta^2 T}{\delta z^2}\right) \tag{9.6}$$

where

r = distance in the radial direction
Θ = angle about the circumference of the cylinder
z = distance along the axis of the cylinder

We will determine thermal conductivity using the probe method. The probe used for this experiment is shown in Fig. 9.1. The probe consists of a line heat source (constantan wire) inside the length of a hypodermic needle and a chromel-constantan thermocouple located midway along the needle. The probe is inserted into the sample whose thermal conductivity is to be determined and a constant current is applied to the heater wire. For the probe method the temperature distribution in the sample at any time, *t*, is described by the differential equation (the heat transfer equation):

$$\frac{\delta T}{\delta t} = \alpha.\left(\frac{\delta^2 T}{\delta r^2} + \frac{1}{r}.\frac{\delta T}{\delta r}\right) \tag{9.7}$$

whose "simplified" solution is:

$$T - T_o = \frac{Q}{4\pi.k}\left[-\delta - \ln\frac{r}{4\alpha.t} + \frac{r^2}{4\alpha.t} - \frac{1}{4}\left(\frac{r^2}{4\alpha.t}\right)^2\right] \tag{9.8}$$

where

Q	=	heating rate/m or ft of heater wire
δ	=	Euler's constant
T_o	=	initial temperature

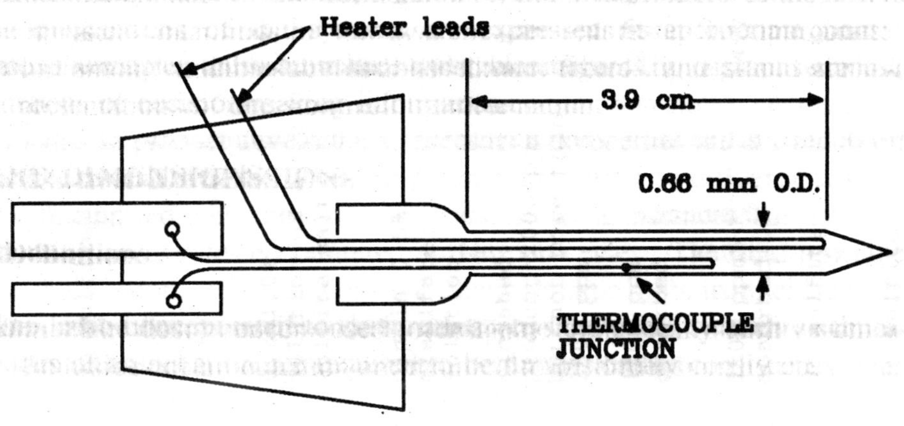

Fig. 9.1 **A schematic of the thermal conductivity probe.**

After a short warmup period, a graph of *T* vs. ln (*t*) yields a straight line. For two different times on the linear portion of the curve, the following relationship holds:

$$T_2 - T_1 = \frac{C.Q}{4\pi.k}\ln\left(\frac{t_2}{t_1}\right) = \frac{C.Q}{4\pi.k}\,2.3\log\left(\frac{t_2}{t_1}\right) \tag{9.9}$$

where C is a calibration constant (probe constant) for the particular probe used. In the design of thermal conductivity probes it is necessary to consider the length to diameter ratio (see Mohsenin, 1980, for further elaboration), and the minimum sample size required to avoid changes in temperature at the sample boundary.

Rearranging equation 9.9, we can obtain expressions for the probe constant C, or the thermal conductivity k. T_2, T_1, t_2, and t_1 are obtained by plotting T vs. ln (t) or log (t) from experimental data (Fig. 9.2). By definition, $Q = I^2R_h$, where

I	=	current, A
R	=	resistance, Ω
h	=	heater

The resistance of the heater wire (R_h) should be measured in Ω/m. The supply voltage should be approximately 10 V D.C., and the current will be measured.

Thus, the theory of the thermal conductivity probe is based on the line heat source method. It utilizes a constant heat source to an infinite solid along a line with infinitesimal diameter, such as a thin resistance wire. The rate of rise in temperature of the sample is a function of the thermal conductivity. The thermal conductivity is calculated using:

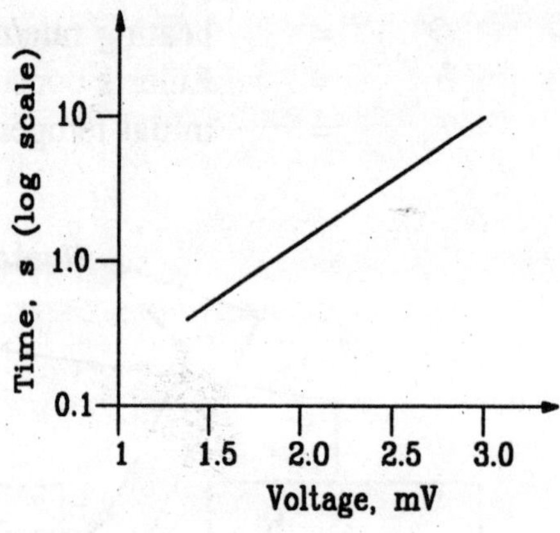

Fig. 9.2 Voltage vs. log (t) for water at 20°C.

$$k = \frac{C.I^2.R_h}{4\pi(T_2 - T_1)} \; 2.3 \log (t_2/t_1) \tag{9.10}$$

9.1.3. Thermal Diffusivity

In this lab we will determine the thermal diffusivity of a food using the method of Dickerson (1965). The apparatus used is shown in Fig. 9.3. In this method the temperature of the sample and the environment (water bath) are increased linearly with time. Thus, after an initial transient period, the temperature difference between the sample and its environment is constant (see Fig. 9.4). This makes the $\delta T/\delta t$ term in equation 9.7 a constant. If the temperature change along the axis of the cylinder containing the sample is eliminated by agitating the water bath, then the terms $\delta^2T/\delta z^2$ and $\delta^2T/\delta\Theta^2 = 0$. Therefore we can write:

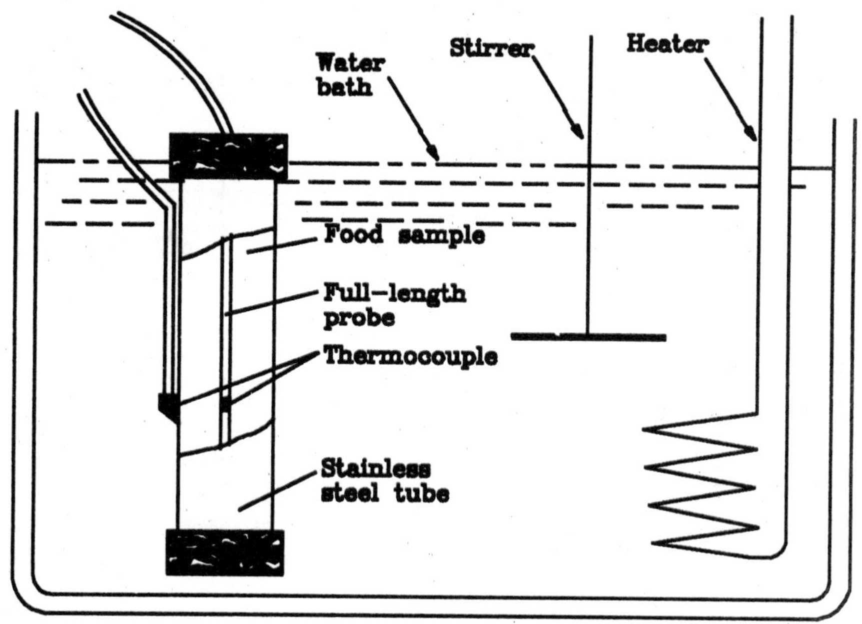

Fig. 9.3 Thermal diffusivity setup.

$$\frac{\delta T}{\delta t} = A = \alpha.\left(\frac{\delta^2 T}{\delta r^2} + \frac{1}{r}.\frac{\delta T}{\delta r}\right) \tag{9.11}$$

or

$$\frac{A}{\alpha} = \frac{\delta^2 T}{\delta r^2} + \frac{1}{r}.\frac{\delta T}{\delta r} \tag{9.12}$$

Now, because the temperature gradient $\delta T/\delta r$ is no longer time dependent but depends only on the radius of the sample, we can solve the differential equation by integrating twice. The solution of equation 9.12 is:

$$T = \frac{A.r^2}{4\alpha} + C_1 \ln(r) + C_2 \tag{9.13}$$

The boundary conditions are:

$$T = A.t = T_R \text{ for } t > 0, r = R$$

$$\frac{dT}{dr} = 0 \quad \text{for } t > 0, r = 0$$

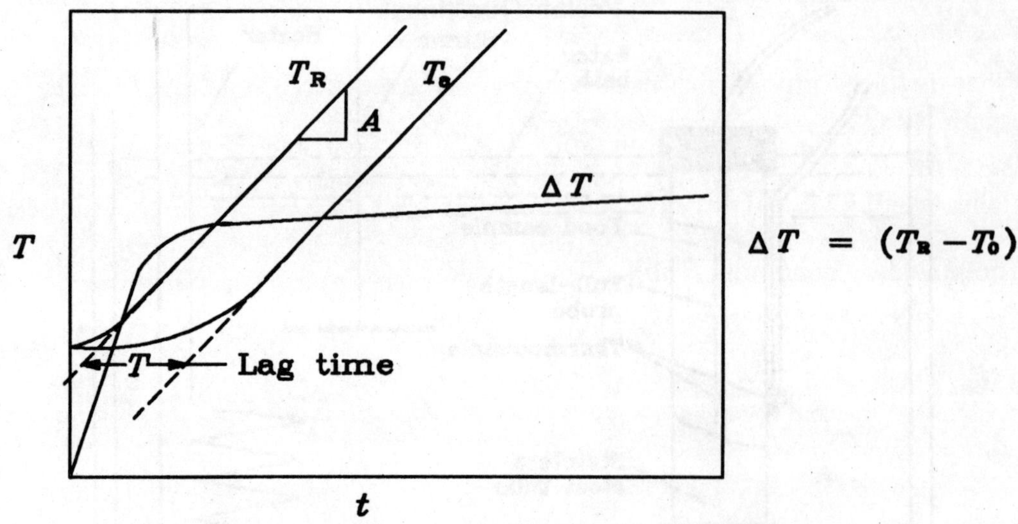

Fig. 9.4 **Data plot and analysis for thermal diffusivity.**

Substituting these boundary conditions into equation 9.13 gives us

$$T_R - T = \frac{A}{4\alpha}(R^2 - r^2) \tag{9.14}$$

At $r = 0$, we get

$$T_R - T_0 = \frac{A.R^2}{4\alpha} \tag{9.15}$$

or

$$\alpha = \frac{A.R^2}{4(T_R - T_0)} \tag{9.16}$$

where

A	=	slope of the heating curve, °C/min
R	=	radius of cylinder, cm
T_R	=	water bath temperature at any time t, °C
T_O	=	cylinder center temperature at t, °C
α	=	thermal diffusivity, cm²/min

This approximation has been shown to be within 5% of the exact analytical solution when $\alpha.t/R^2 \leq 0.55$.

Prelab Questions

Q1. What is the form of the heat equation in three-dimensional rectangular Cartesian coordinates?

Q2. What assumptions are made in reducing equation 9.6 to equation 9.7?

Q3. The boundary condition $dT/dr = 0$ is equivalent to saying that T remains finite as $r \rightarrow 0$. What value of T would we get if this restriction was not imposed?

9.2. OBJECTIVES

The objectives of this laboratory are to:

1. Introduce several differential equations describing heat transfer.

2. Determine experimentally the specific heat, thermal conductivity, and thermal diffusivity of selected food materials.

3. Familiarize you with several methods of estimating the thermal properties of foodstuffs based on composition and structure.

9.3. APPARATUS

9.3.1. Specific Heat

1. Magnetic stirrer (2)
2. Stirring bars (2)
3. Adiabatic flask (2)
4. 250 mL beaker with sample (2)
5. Temperature measuring system
6. Water bath (1)
7. 1000 mL graduated cylinder (1)
8. Electronic balance

9.3.2. Thermal Conductivity

1. Thermal conductivity probe (1)
2. 0 to 20 V D.C. power supply (1)
3. Millivolt recorder (1)
4. Digital voltmeter (1)
5. Sample cell (1)

9.3.3. Thermal Diffusivity

1. Sample tube (2)
2. Water bath with thermograd (1)
3. Micromite or digital voltmeter with electronic ice-point reference, or a temperature recorder.

9.4. PROCEDURE

9.4.1. Specific Heat

1. Using a water bath, heat the food sample to the test temperature plus about 10°C. Allow the sample to equilibrate for 10 min.

2. Heat or cool the flask by adding a charge of water (about 700 mL) that is about 10°C below the test temperature. Weight the water charge and record T_i at 30 s intervals until mixing with the sample. Record the data in Table 9.1.

3. Weight the product and record T_s.

4. Add the product to the flask and record T_m at 30 s intervals until constant.

5. When the effective heat capacity of the flask is unknown, then procedure 1 to 4 can be repeated for a product p (such as water) with a known specific heat. Use Table 9.2 for data recording. Equation 9.1 may be rearranged to solve for the calorimeter constant $C_f W_f$:

$$C_f W_f = \frac{C_c W_c\,(T_m - T_i) + C_p W_p\,(T_m - T_p)}{(T_m - T_i)} \qquad (9.17)$$

6. Plot temperatures (T_s, T_i, and T_m) versus time. This will allow you to extrapolate to determine T_s, T_i, and T_m at the time of mixing. This will allow for heat exchange with the flash environment.

7. To calculate $C_f W_f$, an additional plot of temperatures (T_p, T_i, and T_m) versus time will be required.

9.4.2. Thermal Conductivity

1. Fill the sample cell with glycerol (k = 0.284 W/(m.K)) at 20°C. This will be used as a reference to obtain the probe constant C.

2. Insert probe, turn on recorder (0 to 10 mV, 10 mm/s), or data acquisition system, and heat the probe by applying a constant current for approximately 25 s.

3. Replace the glycerol with the food sample and repeat steps 1 and 2. Record the data in Table 9.3.

4. Using the probe constant C, obtained from the data for glycerol, calculate the thermal conductivity of the sample.

9.4.3. Thermal Diffusivity

1. Load product tightly into sample tubes.

2. Insert the thermocouple and clamp the end plate of the tube. Make sure the surface thermocouple is properly glued to the tube.

3. Place the entire assembly in the water bath. Using the thermograd, heat at a constant rate to satisfy the requirement that $\delta T/\delta t$ is constant.

4. Record the bath, center temperature (T_O), and surface temperature (T_R) at two-minute intervals in Table 9.4 until the surface temperature reaches 80°C.

9.5. RESULTS AND DISCUSSION

9.5.1. Specific Heat

Report your results as shown in the sample data tables provided. Calculate the specific heat, and compare the average value obtained experimentally to those values calculated from the models available in the computer program provided (obtain the required food compositional data from the literature).

9.5.2. Thermal Conductivity

Plot T vs. log (t) (or T vs. t using a semilog paper) for both the glycerol and your food sample (choose the number of points necessary to produce a good fit). Calculate the probe constant and the thermal conductivity of the sample. Using the models provided in a computer program, calculate the expected thermal conductivity of your sample. Answer the following questions as part of your lab report:

1. Is this a steady state or a transient method? What is the difference between the two?

2. At what temperature is the thermal conductivity of the sample being determined?

3. What sources of error may cause the probe constant to be something other than 1.0?

9.5.3. Thermal Diffusivity

Plot T_O, T_R, and $(T_R - T_O)$ against time on the same graph, and fit the data with smooth curves. Determine the slope of the bath heating curve A in °C/min. Calculate the thermal diffusivity using equation 9.16. Again, using a computer program provided, determine the expected value of the thermal diffusivity and discuss any discrepancies.

Answer the following questions as part of your lab report:

1. Is this a steady state or transient method?

2. At what temperature is α being determined?

9.5.4. Additional Questions

1. Why are thermal properties important?

2. Explain the relationship behind specific heat (C_p) and enthalpy (H) conceptually and thermodynamically.

3. What is the most important component of a food with respect to its influence on the thermal conductivity? Why?

4. How may a "heat penetration" test be used to determine the apparent thermal diffusivity of a product at retort or cooling temperature?

5. Why does the center of a container with food not immediately start heating after a container is put into a retort or autoclave?

6. What shape container of the same radius and the same characteristic dimension would heat faster--a slab, a long cylinder, or a sphere? Why?

9.6. ALTERNATE METHOD OF DETERMINING THERMAL DIFFUSIVITY

Thermal diffusivity from the response of a can shaped object to a step change in environment.

9.6.1. Objective

To determine the thermal diffusivity of a food in a can.

9.6.2. Background

The analytical solution to one-dimensional transient heat conduction equation is the sum of a series of terms. After a reasonable time (when, ΔT, change in temperature of the food material is less than 0.5 (T_e - T_O), where T_e is environment temperature and T_O is initial food temperature, one term dominates:

$$\log \Delta T = -t/f + \log j + \log \Delta T_O \tag{9.18}$$

and

$$\alpha = \frac{2.303 r^2}{\beta^2 f}$$

where

α	=	thermal diffusivity, m^2/s
r	=	radius of product in can, m
t	=	time, s
f	=	time for 90% change (one log cycle) on the straight line portion of log (T_e - T) vs. t plot
β	=	shape factor for slab, $\beta = \pi/2$ for cylinder, $\beta = 2.40$ for sphere, $\beta = \pi$ for can, $\beta = [2.4^2 + (r^2/a^2)(\pi/2)^2]^{0.5}$ where a is half the height of the can
ΔT_O	=	(T_e - T_O)
j	=	(T_e - T_a)/(T_e - T_O)
T_a	=	apparent initial food temperature, to be determined from log (T_e - T) vs. t plot

9.6.3. Apparatus

Agitated control temperature water bath (2), vernier caliper, spatula, weight balance, can opener, cans, can sealing machine, Ecklund thermocouple with fittings, and temperature recording system.

9.6.4. Procedure

1. Fill the can 90% full with the food product after installing Ecklund thermocouple assembly from the side of the can at its geometric center. Be sure the gaskets are tightened.

2. Measure and record dimensions and mass of the can.

3. Turn on temperature recording mechanism and measure water bath temperature. Connect the can with the recorder using the imbedded thermocouple and immerse in hot water bath. Use tongs and gloves. Do not transfer by the thermocouple wires. Do not drop.

4. Heat until the product is within 2°C of the bath temperature. Then transfer the container to the cool bath, and turn off the recorder. Record the data in Table 9.5.

5. Upon cooling, weigh can after all food product has been washed out.

9.6.5. Results and Discussion

1. Record the data in the attached data sheet. Ten to 20 temperature data points are needed.

2. Plot $(T_e - T)$ versus time t on a semilog coordinate. Determine f, j, and α.

9.7. COMPUTER PROGRAMS

The computer program "Thermal Properties" is explained in the Chapter on "Computer Programs". This program can be used to compute various thermal properties of the foods based on their compositions and temperatures. Compute these thermal properties of the foods used in this laboratory and compare with the corresponding experimental values.

Table 9.1 Data Sheet to Determine Specific Heat of a Food Sample (C_s)

Food sample information ____________________

Food sample composition ____________________

Food sample mass, W_s ____________________

Charge of water, W_c ____________________

Specific heat of the charge, C_c ____________________

Time, s	Temperature of the Charge, °C (T_c)	Temperature of the Food Sample, °C (T_s)	Temperature of the Mixture, °C (T_m)
0			
30			
60			
90			
120			
150			
180			
210			
240			
270			
300			

Table 9.2 Data Sheet to Determine $C_f W_f$ (Heat Capacity of the Flask)

Product ______________________ Mass, W_p ________________

Charge of water, W_c ______________________

Specific heat of the charge, C_c ______________________

Time, s	Temperature of the Charge, °C (T_c)	Temperature of the Product, °C (T_p)	Temperature of the Mixture, °C (T_m)
0			
30			
60			
90			
120			
150			
180			
210			
240			
270			
300			

Table 9.3 Data Sheet to Determine Thermal Conductivity of a Food Sample

Resistance of the heater wire, R_h ______________________ Ω/m

Applied current, I ______________ A

Initial food sample temperature ______________________ °C

Food sample specification ______________________

Food sample composition ______________________

Correction factor, C ______________________

Data Number	Time, s	Temperature, °C	Log (time) if Required
1			
2			
3			
4			
5			
6			
7			
8			
9			
10			
11			
12			
13			
14			
15			
16			
17			
18			
19			
20			

Table 9.4 Data Table for Thermal Diffusivity Determination

Date ______________________________

Product information ______________________________

	RAW DATA				CALCULATED DATA							
Time from Start	Bath Temp., T_R, °C		Thermal Grad Reading		T_O		T_R - T_O		A		α	
	1	2	1	2	1	2	1	2	1	2	1	2

Table 9.5 Data Sheet to Determine Thermal Diffusivity

Can dimensions: length ____________________ cm, diameter ____________________ cm

Can mass: filled ____________________ g, empty ____________________ g

Mass of food in the can: ____________________ g

Food product information:____

____

____

Food product composition: ____

Hot water bath temperature, T_e ____________________ °C

Initial food product temperature, T_O ____________________ °C

Time, min	Product Center Temperature (T), °C	(T_e - T), °C
0		
2		
5		
10		
15		
20		
25		
30		
35		
40		
45		
50		

Prelab Questions for Laboratory 9

NAME:____________________________________ **DATE:**________________________

ANSWER PRELAB QUESTIONS ON THIS SHEET

Prelab Questions for Laboratory 1

NAME ______________________ **DATE** ______________________

ANSWER PRELAB QUESTIONS ON THIS SHEET

Laboratory 10

SURFACE HEAT TRANSFER COEFFICIENT

SUMMARY

This lab is designed to measure surface heat transfer coefficient of a can shaped object heated by water. This will provide the concept related to the influence of heat transfer coefficient on the response of a product to a thermal process.

10.1. BACKGROUND

The heat transfer coefficient becomes important when the resistance of the surface film on the product (or container) surface is large relative to the product thermal resistance. An object that is a very good conductor of heat, compared to the external conductance, is at nearly the same temperature, T, throughout. An energy balance on this object for a time interval Δt is given by:

Heat stored = Heat added

$$V.\rho.C_p.\ \Delta T\ = h.A.(T_e - T)\ \Delta t \qquad (10.1)$$

where

V = object volume
ρ = object density
C_p = object-specific heat
ΔT = temperature change
h = heat transfer coefficient
A = surface area of the object
T_e = environment temperature
T = temperature of the object
Δt = time interval

The Biot number is the ratio of conductive resistance in the solid to convective resistance in the fluid. It is defined by Bi = $h.L/k$, where L is the characteristic length and k is the thermal conductivity of the solid material. The L for sphere and cylinder is the radius and for flat plat is its thickness or half the thickness if the heat transfer is taking place from both sides of the plate. When Bi $\rightarrow$ 0, conductive resistance will tend to be zero, and when Bi $\rightarrow \infty$, convective resistance will be equal to zero or h is very large. When Bi approaches zero, the solid is practically isothermal and the temperature varies most in the fluid. As Bi approaches infinity, the fluid is nearly isothermal, and the temperature differences occur mainly in the solid.

Thus, when a body has negligible internal (conductive) thermal resistance, the temperature gradients inside the body are much smaller then those occurring in the surrounding fluid. For an irregularly shaped body the characteristic length L is defined as the volume of the body divided by its surface area. When Bi is much less than 2, the change in stored energy in the solid should be equal to the heat transfer rate from the surface by convective mode. If the Biot number is less than 0.10, the error in the temperature history is less than 5% when calculated by using the above approach. As Biot number becomes smaller, the accuracy is increased.

A number of correlations have been developed to calculate h by knowing the Reynolds (Re) and Prandtl (Pr) numbers (see any heat transfer book such as Kreith and Black, 1980 for details). Thus, if the object volume, area and specific heat are known, a heating or cooling experiment at constant environment temperature T_e can be used to determine h. An aluminum cylinder (dimensionally similar to a can), even when placed in a heated water bath is sufficiently small and is of such high thermal conductivity that its Biot number is low and we can assume the entire cylinder at a single temperature at any given moment.

Rearranging equation 10.1:

$$\frac{\Delta T}{\Delta t} = \frac{h.A.}{V.\rho.C_p}(T_e - T) \qquad (10.2)$$

Thus, the plot of $\Delta T/\Delta t$ versus $(T_e - T)$ should be a straight line with slope = $h.A./(V.\rho.C_p)$, if thermal properties h and C_p are constant.

Also, rearranging equation 10.1 again gives:

$$\frac{\Delta T}{T_e - T} = \frac{h.A}{V.\rho.C_p}\Delta t \qquad (10.3)$$

or for an infinitesimal change in time, this can be written as

$$\frac{dT}{T_e - T} = \frac{h.A}{V.\rho.C_p}\,\mathrm{d}t \qquad (10.4)$$

and integrated equation 10.4 to obtain

$$\ln\left(\frac{T_e - T}{T_e - T_0}\right) = \frac{-h.A}{V.\rho.C_p}.t \qquad (10.5)$$

where T_O is the initial temperature of the cylinder, and $(T_e - T)/(T_e - T_O)$ is the temperature ratio TR.

Rewriting equation 10.5 we get

$$\log(TR) = \frac{-h.A}{2.3V.\rho.C_p} \cdot t \qquad (10.6)$$

Thus, the plot of log (TR) vs. t should be a straight line with slope = - $h.A./(2.3V.\rho.C_p)$.

Prelab Questions

Q1. Define surface heat transfer coefficient conceptually and mathematically.

Q2. What are the factors affecting *h*?

Q3. Provide three models to predict *h* for regular geometries.

Q4. When can *h* be neglected in heat transfer calculations?

10.2. OBJECTIVES

1. To determine the surface heat transfer coefficient of a small cylinder immersed in water.

2. To verify the first order relationship between temperature and time under conditions of a small Biot number, $h.R/k$.

10.3. APPARATUS

1. An aluminum cylinder, with an embedded thermocouple at the geometric center of the cylinder.
2. A temperature recording system.
3. A water bath.
4. A vernier caliper.
5. A weighing scale.

10.4. PROCEDURE

1. Measure and record dimensions of the cylinder.

2. Connect the thermocouples of the cylinder and water bath to the temperature recording mechanism. Record the initial temperatures after turning on the recorder.

3. Transfer the aluminum cylinder into the water bath. Make the transfer with tongs. Do not grab the wire. Do not drop the cylinder. Record the temperatures at 5 s intervals in Table 10.1.

4. Record the temperature until the temperature change is small.

5. Stop the recorder and return the cylinder to a cold water bath.

10.5. RESULTS AND DISCUSSION

1. Tabulate the time-temperature data in the attached data sheet. About 10 to 15 data points are required.

2. Convert temperature-time data to determine $\Delta T/\Delta t$ and TR.

3. Plot $\Delta T/\Delta t$ vs. $(T_e - T)$ on a linear graph paper.

4. Plot log (TR) vs. time t on a semilog graph paper (2 cycle).

5. Based on your best straight line through the data points and estimate the slope and h values.

6. Calculate the approximate value of the Biot number for the cylinder in water to check the validity of the assumption of uniform cylinder temperature during heating or cooling in water.

Table 10.1 Data Sheet for Laboratory 10

Cylinder length ______________ cm, diameter ____________ cm, mass ____________ g

Cylinder volume ______________ m^3, density ____________ kg/m^3

Cylinder surface area ______________ m^2, Biot number ____________________

Initial temperature of the cylinder ____________________ °C

Initial temperature of the water bath ____________________ °C

Data Point	Time, s	Temperature, °C		$T_e - T$	$\Delta T/\Delta t$	TR
		Bath, T_e	Cylinder, T			

Prelab Questions for Laboratory 10

NAME:______________________________ **DATE:**____________________

ANSWER PRELAB QUESTIONS ON THIS SHEET

Laboratory 11

THERMODYNAMICS OF FOOD FREEZING

SUMMARY

The goals of this laboratory exercise are to compare experimentally determined freezing times with those obtained from predictive equations and to determine the effects of packaging on freezing times.

11.1. BACKGROUND

11.1.1. Freezing Process

Freezing is a common method for long-term preservation of foods and other biomaterials. It involves crystallization of most of the water and some of the solutes by reducing the product temperature to -18°C (0°F) or lower. The beneficial effects of preservation by freezing are attributed to a significant reduction in the rate of chemical, biochemical, and microbiological activities in the product. Additionally, crystallization of water into ice changes the availability and mobility of water to participate in any of these reactions.

In freezing operations, the product to be frozen is exposed to a temperature much lower than the desired final product temperature. This results in the removal of sensible heat followed by latent heat from the product and crystallization of water into ice. The temperature at which ice formation begins is called the initial freezing point of the product. The freezing and crystallization process taking place within a food system are different from those occurring when pure water is frozen. This is illustrated in Fig. 11.1 by the freezing curves for water and an ideal binary aqueous solution. On cooling the ideal solution below its initial freezing point, point 1, pure ice is formed and therefore solutes in solution become more and more concentrated, and the freezing point is depressed further and further.

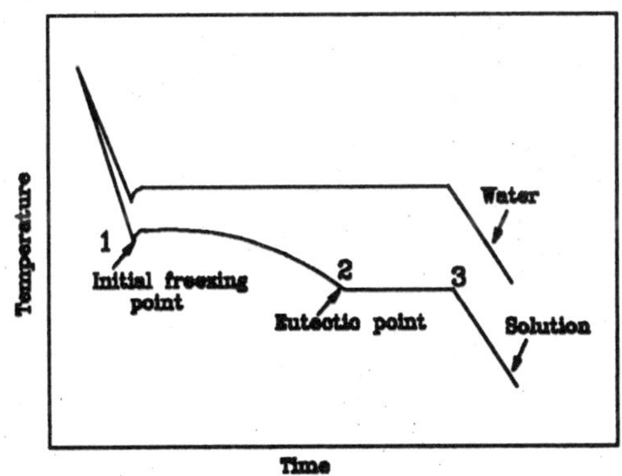

Fig. 11.1 A comparison of freezing curves for pure water and aqueous solution containing one solute.

As point 2 is approached, the liquid phase becomes super-saturated and the solute crystallizes out. At point 2, the crystallized solute exists in equilibrium with ice and the unfrozen, concentrated phase. The temperature at this equilibrium condition is known as the eutectic temperature or eutectic point of the solute. Further removal of heat beyond the eutectic point results in simultaneous crystallization of water and solute in a constant ratio and the composition of the solution remains constant. The crystallization process is completed at point 3; subsequent cooling results in decrease in product temperature. It must be recognized that the freezing process becomes much more complex when foods are involved. This is to be expected because of various types and amounts of dissolved solutes that exist in food systems, each with different eutectic points and heats of crystallization.

Food engineers must deal with two major aspects of freezing: (1) designing a good freezing process along with the selection of proper equipment and (2) estimating the refrigeration requirements for freezing systems. These tasks require a knowledge of the product freezing rate, defined as a time derivative of ice front movement or temperature. The rate of freezing also affects frozen product quality, depending on the food commodity. At slow rates of freezing, few crystallization centers are formed and the ice crystals grow to relatively large sizes. This may lead to mechanical damage of some frozen food products and decrease their textural quality on thawing. At high freezing rates, on the other hand, the number of ice crystal centers increases and their size decreases. Rapid freezing, although desirable for some products, may not justify the added costs in other situations because of either the size and configuration of the products or its structural integrity.

11.1.2. Estimation of Freezing Time

As previously mentioned knowledge of the freezing rate and therefore the freezing time is the most fundamental information needed by an engineer to select and design a freezing process, and to establish system capacity requirements. The freezing time requirements depend on such factors as the size and shape of the product, thermal conductivity of the product, initial and final temperatures, change in enthalpy, surface heat transfer coefficient of the system, and temperature of the refrigerating medium. Because of these varying factors, calculations of freezing times become difficult. For a particular product, however, the freezing time can be experimentally determined by measuring the product temperature at its slowest cooling point.

Freezing times of foods can be predicted by either analytical methods or by numerical methods. The latter solve the three- dimensional heat conduction equation with the appropriate initial and boundary conditions and thermophysical properties. A complete analytical solution for predicting the freezing (or thawing) times of food products require tedious calculations. Various simplified models for predicting the freezing times have been proposed by Plank (1913), Hohner and Heldman (1970), Cleland and Earle (1979), Hayakawa, Nonino, and Succar (1983), and Cleland and Earle (1987), among others. Plank's equation and Cleland and Earle (1979) equations will be used in this exercise.

Plank's model has been used for its simplicity and convenience. Its basic assumptions include (1) commencement of the freezing process with all of the food unfrozen but at its initial freezing temperature, (2) equality of the initial and final temperatures with the initial freezing temperature, (3) homogeneity and isotropy of the product that provides invariability of its thermophysical properties, and (4) steady-state heat transfer in the frozen layer of the product.

The case of one-dimensional freezing of an infinite product slab of thickness a is being cooled by convection in an environment of constant temperature T_m as illustrated in Fig 11.2. The frozen layer grows with time and at any time t, a thickness x of frozen layer is formed on both sides. The initial freezing temperature is constant at T_f. The unfrozen center is also at T_f.

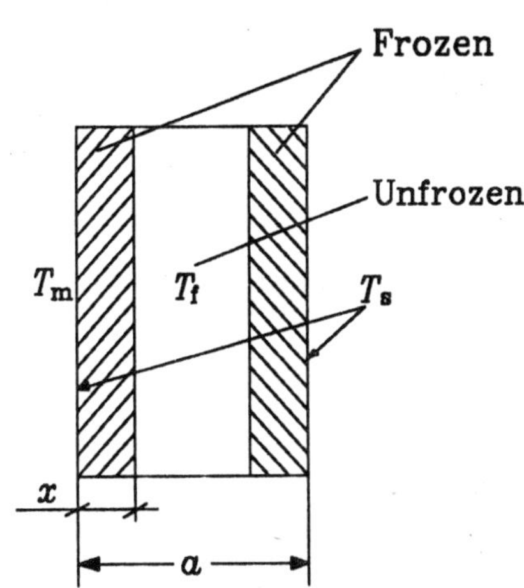

Fig. 11.2 Schematic illustration of one-dimentional freezing of a product section used to derive Plank's equation.

If, in time dt, a layer dx thick of the product freezes, then the rate of heat generation becomes

$$q = A \lambda \rho \ (dx/dt). \qquad (11.1)$$

where A is the surface area, λ is the latent heat of freezing, and ρ is the density of the frozen product. At steady state, the heat given off at the freezing front of the product must be removed by conduction through the frozen layer of thickness x followed by its removal by convection at the outside surface. The equation describing heat transfer by conduction is

$$q = \frac{k.A \ (T_f - T_s)}{x} \qquad (11.2)$$

where k is the thermal conductivity of the frozen product and T_s is the surface temperature.

The second equation describing the convective heat transfer at the surface is

$$q = h.A \ (T_m - T_s) \qquad (11.3)$$

where h is the convective heat transfer coefficient at the product surface and T_m is the medium temperature. By eliminating the surface temperature variable T_s, equations 11.2 and 11.3 can be combined to describe heat transfer in series as follows:

$$q = \frac{A(T_f - T_m)}{x/k + 1/h} \qquad (11.4)$$

Equating equation 11.1 to 11.4, rearranging and integrating between $t = 0$ and $x = 0$ to $t = t_f$ and $x = a/2$ gives

$$(T_f - T_m) \int_0^{t_f} dt = \lambda.\rho \int_0^{a/2} \left(\frac{x}{k} + \frac{1}{h}\right) dx \qquad (11.5)$$

Integrating and solving for freezing time t_f yields Plank's equation:

$$t_f = \frac{\lambda.\rho}{T_f - T_m} \left(\frac{a}{2h} + \frac{a^2}{8k}\right) \qquad (11.6)$$

The general form of the Plank's equation for other shapes is

$$t_f = \frac{\lambda.\rho}{T_f - T_m} \left(\frac{P.a}{h} + \frac{R.a^2}{k}\right) \qquad (11.7)$$

where a is the total thickness of an infinite slab, diameter of a sphere, diameter of a long cylinder, or the smallest dimension of a rectangular brick. P and R are constants determined by the geometry of the product being frozen. These constants for various geometries are given below:

Infinite slabs:	$P = 1/2$	$R = 1/8$
Infinite cylinders:	$P = 1/4$	$R = 1/16$
Sphere:	$P = 1/6$	$R = 1/24$

For brick-shaped products, P and R are obtained from Fig. 11.3 prepared by Ede (1949) where β_1 and β_2 are the ratios of the two longest sides divided by the shortest. This would provide a value for either P or R. The other value for P or R is obtained by interchanging β_1 and β_2.

Despite its simplifying assumptions, Plank's equation gives satisfactory results as long as the product is initially at its freezing temperature. Most other available analytical models for freezing time calculation are modifications of Plank's equation with emphasis on developments to overcome limitations of the original equation. Such a modification has been proposed by Nagoaka, Takagi, and Hotani (1955), which accounts for the initial product temperature above its freezing temperature and the final product temperature below the freezing point. Their modified equation is

$$t_f = [C_u(T_i - T_f) + X_w\lambda + C_f(T_f - T)]\,[1 + 0.008\,(T_i - T_f)] \left[\frac{\rho}{T_f - T_m}\right] \left[\frac{Pa}{h} + \frac{Ra^2}{k}\right] \qquad (11.8)$$

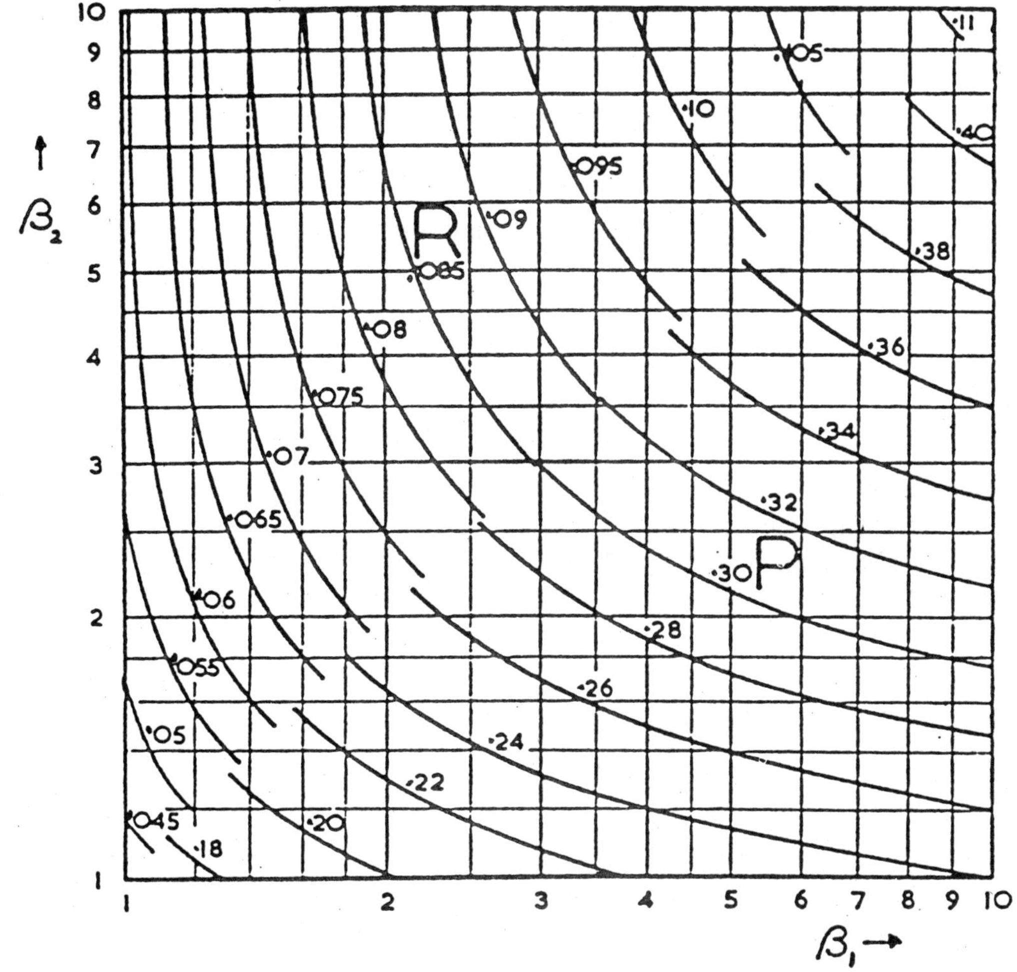

Fig. 11.3. Values of *P* and *R* for a brick-shaped product (based on Ede, 1949).

where

T_f = freezing point of material
X_w = water content (mass fraction), wet basis
T_i = initial product temperature
T_m = freezing medium temperature
T = final product temperature
C_u = specific heat of unfrozen material
C_f = specific heat of frozen material
λ = latent heat of fusion
ρ = density of frozen material

A factor of 0.0045 will replace 0.008 if English units are used. When the product to be frozen is contained in a package, the convective heat transfer coefficient h in equations 11.7 and 11.8 is replaced with an overall heat transfer coefficient U to account for the resistance to heat flow offered by the packaging material. For a packaged system, the value of U is defined as

$$U = \frac{1}{x_p/k_p + 1/h} \tag{11.9}$$

where x_p is the thickness and k_p is the thermal conductivity of the packaging material.

Cleland and Earle (1979) modified Plank's equation by rewriting it in a dimensionless form:

$$N_{FO} = \frac{P}{N_{Bi}\ N_{St}} + \frac{R}{N_{St}} \tag{11.10}$$

where

N_{Fo} = Fourier number = $\alpha.t/a^2$
N_{Bi} = Biot number = $h.a/k$
N_{St} = Stefan number = $C_f(T_f - T_m)/\lambda$
N_{Pk} = Plank's number = $C_u(T_i - T_f)/\lambda$
α = thermal diffusivity

P and R were calculated by the following equations:

Slab

$$P = 0.5072 + 0.2018N_{Pk} + N_{St}(0.3224N_{Pk} + 0.0105/N_{Bi} + 0.0681) \tag{11.11}$$

$$R = 0.1684 + N_{St}\ (0.2740N_{Pk} + 0.0135) \tag{11.12}$$

Cylinder

$$P = 0.3751 + 0.0999N_{Pk} + N_{St}(0.4008N_{Pk} + 0.0710/N_{Bi} - 0.5865) \tag{11.13}$$

$$R = 0.0133 + N_{St}\ (0.0415N_{Pk} + 0.3957) \tag{11.14}$$

Sphere

$$P = 0.1084 + 0.0924N_{Pk} + N_{St}(0.2310N_{Pk} - 0.3114/N_{Bi} + 0.6739) \tag{11.15}$$

$$R = 0.0784 + N_{St}\ (0.0386N_{Pk} - 0.1694) \tag{11.16}$$

The values of thermophysical properties of a few selected materials are shown in Tables 11.1 to 11.4. The *ASHRAE Handbook of Fundamentals* is a good source of data on numerous food products.

Prelab Questions

Q1. What is an eutectic point?

Q2. What is the latent heat of freezing for water in English Engineering units?

Q3. What two basic assumptions were made in the Plank equation?

Q4. What modifications of the Plank equation are found in Nagaoka's equation for freezing time?

Q5. What is the role of the constants *P* and *R* in the equations predicting freezing time? List some foods that could be considered slabs, cylinders, and spheres.

Q6. What physical properties, including dimensions, of a product must be known to use Plank's and Nagaoka's equations?

11.1.3. Sample Problem

A 0.25 m x 0.50 m x 0.75 m block of lean beef initially at 10°C is to be frozen by liquid immersion freezing in Freon-12 to a final temperature of -12°C. Compute the freezing time for the following cases: (1) meat block is unpackaged, (2) meat block is packaged in 1.0 mm thick cardboard.

Solution

The initial and final temperatures are different and therefore the use of Nagaoka et al. (1955) modification is more appropriate (equation 11.8). Next, the physical and thermal properties of the product being frozen are obtained.

X_w = 0.68
C_u = 3.5 kJ/(kg.K) (Table 11.1)
T_i = 10°C
T_f = -1.7°C (Table 11.1)
λ = 332.7 kJ/kg (Table 11.3)
C_f = 2.05 kJ/(kg.K) (for ice)
T = -12°C
ρ = 1000 kg/m^3 (assumed)
T_m = -29.8 °C (R-12 at 1 atm)
h = 568 W/(m^2.K) (Table 11.4)
k = 1.1 W/(m.K) (for ice)

To obtain the shape factors *P* and *R*, we first calculate the values of β_1 and β_2.

$$\beta_1 = \frac{0.50}{0.25} = 2 \quad and \quad \beta_2 = \frac{0.75}{0.25} = 3$$

From Fig. 11.3, the values obtained are $P = 0.275$ and $R = 0.078$.

1. Substituting the foregoing values into equation 11.7 gives

$$t_f = [3.5(10 + 1.7) + 0.68(332.7) + 2.05(-1.7 + 12)] \cdot [1 + 0.008(10 + 12)] [\frac{1050}{-1.7 + 29.8}] [\frac{(0.275)(0.25)}{0.568} + \frac{(0.078)(0.25)^2}{1.1 \times 10^{-3}}]$$

$= 57679.36$ s $= 16.02$ h

2. When the meat is packaged, h is replaced with U in equation 11.7. The value of U is obtained from the following data.

$x_p = 1 \times 10^{-3}$ m
$k_p = 0.04$ W/(m.K) (Table 11.3)
Then,

$$U = \frac{1}{\frac{0.001}{0.04} + \frac{1}{568}} = 37.37\ W/(m^2.K)$$

Using the foregoing value of the overall heat transfer coefficient U for h in equation 11.7 along with the properties data previously listed gives $t_f = 74250.7$ s $= 20.63$ h. Thus, the packaging material significantly increased the freezing time because of its insulating properties effects.

11.2. OBJECTIVES

The objectives of this laboratory exercise are to:

1. Compare experimentally determined freezing times with those obtained from predictive equations.

2. Determine the effect of packaging on freezing time.

11.3. APPARATUS

1. A liquid nitrogen, R-12 refrigerant, or refrigerated water bath at -30°C or blast freezer

2. Thermocouples and data logger

3. Polyethylene shrink film, vacuum heat sealer, and hot-air blower

4. Food products for freezing such as apples, banana, frankfurters, beef cubes, and concentrated juices

11.4 PROCEDURE

Apples and various fruit juices will be used to study the effects of packaging on freezing rate and the accuracy of Plank's and Nagaoka's equations.

1. Select two apples as identical as possible for the experiment. Measure the diameter of each, and obtain a density for the apple tissue.

2. Insert and secure a thermocouple near the geometric center of each apple, then shrink wrap one apple in a plastic film.

3. Connect the liquid nitrogen source to the freezer.

4. Connect thermocouples to the recorder and initialize recorder for the thermocouples, data mode, and interval.

5. Record initial products and bath temperatures in Table 11.5.

6. Place products in freezer and start freezing process.

7. Freeze apples to -10°C.

8. Repeat steps 4 to 7 with cans of juice. Freeze juice to -10°C. Obtain a density for the juice.

11.5. RESULTS AND DISCUSSION

1. Plot temperature vs. time for each of the products and treatments used for freezing.

2. Record the thermophysical property data and experimental freezing times on the data sheet, Table 11.5. For computation of freezing times, use the equation of Nagaoka et al. (1955) for the apples, and Plank's equation for the juice. The freezing time t_f for apples is defined as the time from initial temperature to -10°C. For juice, it is defined as the time at the initial freezing point (latent heat removal period). Use computer program "Freezing". Record freezing times obtained in Table 11.5. Show at least one sample calculation.

3. Discuss the results obtained including the effect of packaging and comparison of freezing times in question 2. Identify the most likely sources of error.

4. Using your experimental freezing times, calculate the convective heat transfer coefficient h. Use computer program "Freezing".

5. Plot enthalpy versus temperature for the following products (Table A.11, pp.404, 405, Heldman and Singh, 1981):
   - Fresh lean beef
   - Dried lean beef
   - Carrots
   - Strawberries
   - Peaches

On the same graph, plot enthalpy vs. temperature for pure water starting with zero enthalpy at -40°C. Explain any similarities and/or differences in the curves.

Answer these questions as part of your lab report:

1. Why is Plank's equation used to calculate freezing times for the juice and Nagaoka's equation used for the apples?

2. What effect would you expect due to any trapped air in the packaged products to have on these results for freezing time?

3. Discuss the validity of using mean specific heat to calculate changes in enthalpy and potential problems in calculating refrigeration requirements for food products based on the curves in 5.

Table 11.1 Thermal Properties of Food Products

Product	% Water, (wet basis)	Specific Heat,[a] kJ/(kg.K), (BTU/(lb°F))	Specific Heat,[b] kJ/(kg.K), (BTU/(lb°F))	Average Freezing Point, °C, (°F)
Beef, Lean	68	3.50 (0.84)	-1.7 (28.9)	
Franks	60	3.60 (0.86)	2.34 (0.56)	-1.7 (28.9)
Carrots	88	3.77 (0.90)	1.93 (0.46)	-1.4 (29.5)
Apples	84	3.60 (0.86)	1.84 (0.44)	-2.0 (28.4)
Orange Juice	86	3.89 (0.93)	1.93 (0.46)	-1.2 (29.8)
Apple Juice	87	3.85 (0.92)	1.93 (0.46)	-1.4 (29.4)

[a]Above Freezing, [b]Below Freezing

Table 11.2 Thermal Conductivity of Materials

	k, BTU/(h.ft.°F)	k, W/(m.K)
Cardboard	0.040	0.06
Plastic	0.150	0.26
Tin	35.240	61.00
Steel (1% C)	26.000	45.00
Air (1 atm)		
-45°C (-50°F)	1.181	2.04
-18°C (0°F)	1.311	2.27
10°C (50°F)	1.436	2.49

Table 11.3 Enthalpy of Pure Water

T,°C	H, kJ/kg
-40.00	0.00
-28.88	20.58
-20.55	36.58
-8.88	59.86
0.00 (ice)	78.29
0.00 (liquid)	411.00
10.00	453.47

Table 11.4 Heat Transfer Coefficient for Various Freezing Conditions

Freezing Condition	W/(m^2.K)	BTU/(h.ft^2.°F)
Still air freezing (no radiation)	5.7	1
Air blast freezing (500 ft/min)	17.0	3
Air blast freezing (1000 ft/min)	28.4	5
Liquid immersion freezing	568.0	100

Table 11.5 Data Sheet for Freezing Times

DATE ______________________ Group ______________________

Refrigeration system ______________________________

Property Data	Product #1	Product #2
Dimensions		
Initial freezing point, T_f		
Water content, X_w		
Initial product temperature, T_i		
Freezing medium temperature, T_m		
Final product temperature, T		
Specific heat of unfrozen material, C_u		
Specific heat of frozen material, C_f		
Latent heat of fusion, λ		
Density of frozen product, ρ		
Conductivity of frozen material, k		
Conductivity of packaging material, k_p		
Convective heat transfer coefficient, h		
Shape factors		
P		
R		
Computed freezing times		
Packaged		
Unpackaged		
Experimental freezing times		
Packaged		
Unpackaged		

Prelab Questions for Laboratory 11

NAME:______________________________ **DATE:**____________________

ANSWER PRELAB QUESTIONS ON THIS SHEET

Laboratory 12

COLLIGATIVE PROPERTIES OF FOODS

SUMMARY

This lab is to determine the colligative properties of foods that depend primarily on the concentration of dissolved solutes. These properties include freezing point depression, boiling point depression, boiling point elevation, and osmotic pressure.

12.1. BACKGROUND

The colligative properties, such as freezing point depression, boiling point elevation, and osmotic pressure, depend primarily on the concentration of dissolved solutes because the probability of the solvent changing phase is related to the relative solvent concentration. The ratio between "free" or "available" concentrat-ion (X_a) and the total concentration (X_c) is activity coefficient (γ).

12.1.1. Freezing Point Depression

The shift in the vapor pressure versus temperature curve leads to the Van't Hoff expression:

$$T_O - T_f = \Delta T_f = \frac{-R.T_f.T_O \ln(x_s)}{\Delta H_f} \tag{12.1}$$

where

T_O	=	freezing point of pure solvent in absolute temperature
T_f	=	freezing point of the solution in absolute temperature
R	=	gas constant
X_s	=	mole fraction of solute
ΔH_f	=	molar heat of fusion of the solvent

Equation 12.1 can be rewritten as:

$$\ln (X_s) = \frac{-\Delta H_f \, (T_o - T_f)}{R.T_f.T_o} = \frac{\Delta H_f}{R}\left(\frac{1}{T_o} - \frac{1}{T_f}\right) \tag{12.1}$$

For a very dilute solution, equation 12.2 reduces to:

$$\Delta T_f = \frac{R.T_o^2.M_s.m_s}{1000\Delta H_{f,o}} = \frac{R.T_o^2.m_s}{1000\Delta L_f} = K_f.m_s \tag{12.3}$$

where

M_s = molecular weight of solvent
m_s = molality of solvent
ΔL_f = mass heat of solvent fusion
K_f = molal freezing point depression constant

The depression of freezing point (Fig. 12.1) leads to the requirement for lower storage temperature for frozen foods. Materials such as salts or ethylene glycol are used to reduce freezing points of process coolants and in brine spray freezing of foods. Freezing point depression can be used to check for watering of milk or salt levels of blood.

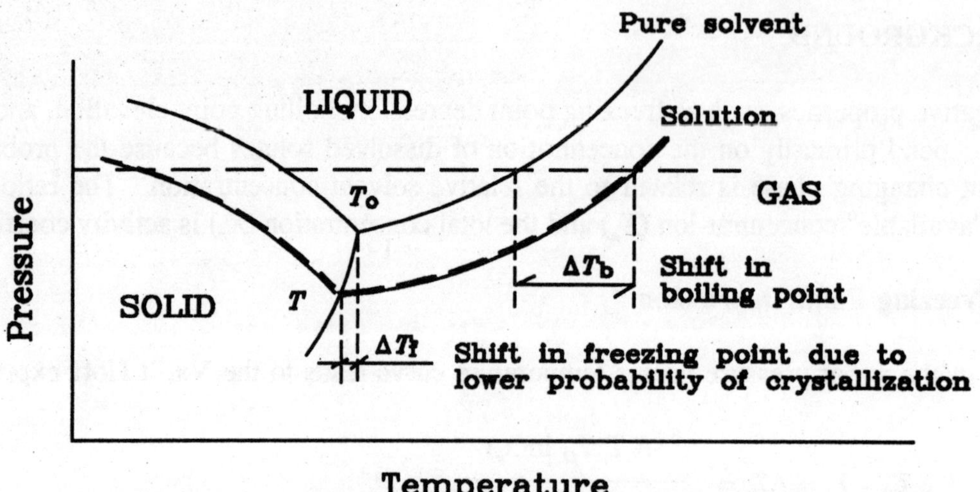

Fig. 12.1 **Pressure–temperature showing freezing and boiling point depression.**

12.1.2. Boiling Point Elevation

The boiling point elevation associated with the addition of a solute is due to the depression of the vapor pressure due to the lowered probability of vaporization and the subsequent increase in thermal energy level required to increase the partial pressure to the total pressure of the surroundings. The shift in vapor pressure follows Raoult's law for dilute solutions ($P = X_s.P_o$), and Henry's law for more concentrated solutions ($P = \gamma.X_s.P_o$) where P is the partial pressure of solution and P_o is the partial pressure of pure solvent.

The log (P) vs. $1/T$ plot (Fig. 12.2) gives straight lines for a moderate range of many solutes.

The equation for this line is known as the Clausius-Clapeyron equation:

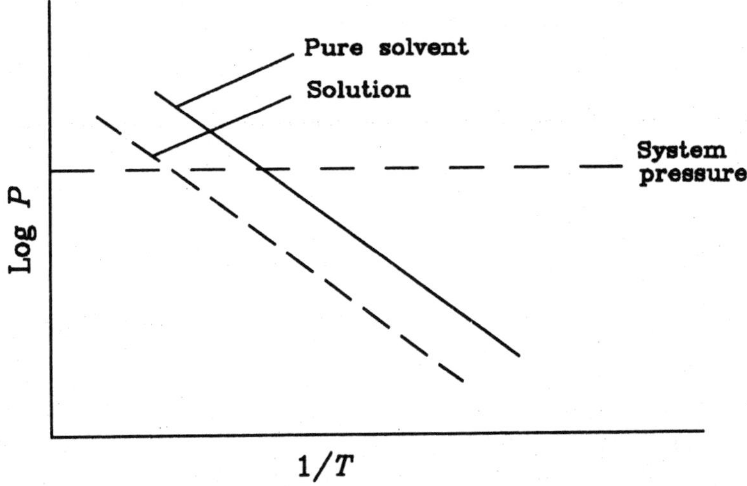

Fig. 12.2 Clausius–Clapeyron diagram.

$$2.3 \log (P/P_o) = \frac{\Delta H_V}{R}\left(\frac{1}{T} - \frac{1}{T_o}\right) \tag{12.4}$$

where ΔH_v is the molal heat of solvent vaporization. The boiling point expressions are of the same form as those for freezing (equations 12.1 to 12.3):

$$T - T_O = \Delta T_b = \frac{R.T.T_O}{\Delta H_v} \ln(X_s) \tag{12.5}$$

and for dilute solutions

$$\Delta T_b = \frac{R.T_o^2.M_s.m_s}{1000\Delta H_{v,o}} = \frac{R.T_o^2.m_s}{1000L_v} = K_B.m_s \tag{12.6}$$

where L_v is the mass heat of solvent vaporization and K_B is a molal boiling point elevation constant. Table 12.1 provides K_B for a few liquids.

12.1.3. Basis for Estimation of Activity Coefficient

We can estimate the activity coefficients for food solutions from freezing point depression or boiling point elevation data.

Table 12.1 Some Molal Boiling Point Elevation Constants (K_B) at the Atmospheric Pressure

Solvent	Normal Boiling Point, °C	K_B, °C/mol
Water	100.0	0.513
Benzene	80.2	2.530
Ethanol	78.3	1.220
Methanol	64.7	2.020

Sugars in very dilute solutions are nearly ideal so their freezing points may be used to calibrate an equipment to measure freezing point (cryoscope). A series of dilutions of a sugar can be made of various molalities. Their freezing point depressions can be estimated from equation 12.3 taking $K_f = 1.86$°C/mol for water. The measured and calculated values of ΔT_f can be used to get a calibration curve. This calibration curve is used to correct subsequent measurements.

The activity coefficient of a liquid food can be determined from the freezing points of several dilutions of the food using $\gamma = \Delta T_f/(K_f.m_s)$, where ΔT_f are the values corrected by using the calibration curve. Figure 12.3 shows the plot of γ versus log (m_s).

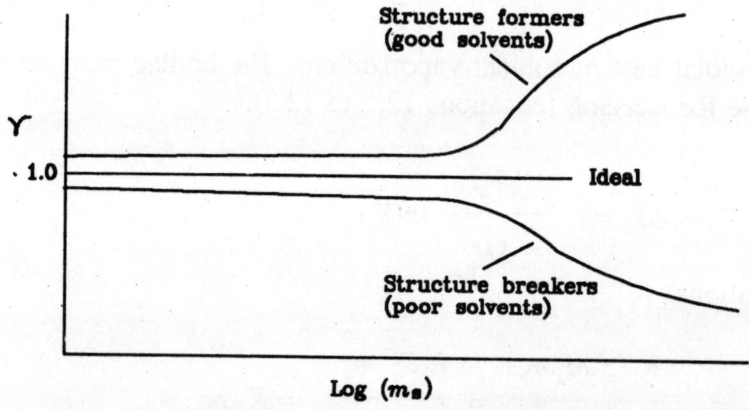

Fig. 12.3 **Relationship between γ and log (m_s).**

12.1.4. Osmotic Pressure π

It is the pressure of a concentrated solution in equilibrium with pure solvent at the same system temperature and pressure. The osmotic pressure is elevated because solvent has moved across the semipermeable membrane barrier between them until its effective concentration is the same on both sides of the barrier. The relationship between osmotic pressure and concentration, C, for an ideal solution is

$$\pi = C.R.T \tag{12.7}$$

The osmotic pressure is used in dialysis to remove small molecular weight components with coarse membranes, and in reverse osmosis with relative tight membranes and added pressure.

Prelab Questions

Q1. Give a conceptual view of why the addition of a solute lowers the vapor pressure.

Q2. Why is boiling point elevation important in determining evaporator capacity?

Q3. How could you use a freezing point depression to monitor the solids concentration of a beverage?

Q4. What is the basis for osmotic pressure?

Q5. What is meant by a good versus a poor solvent?

Q6. The freezing point depressions of three fresh orange juices are as follows:

Juice:	*A*	*B*	Standard
ΔT_f:	0.12°C	0.18°C	0.15°C

What do you believe are the reasons for the behavior of samples A and B, and what are the implications for the production of frozen concentrate?

12.2. EXAMPLES

Calculate colligative properties for a 10% maltose solution assuming ideal behavior (molecular weight = 360.31), K_f for water = 1.86°C/mol, K_b for water = 0.51°C/mol.

<u>Solution</u>

(a) Molality of maltose (m_s)
Maltose in 1000 g water = 10.(1000)/90 = 111.1 g

$$m_s = \frac{111.1 \text{ g}}{360.31 \text{ g/mol}} = 0.3084 \text{ mol}$$

(b) $\Delta T_f = K_f.m_s = (1.86°\text{C/mol})(0.3084 \text{ mol}) = 0.\underline{574}°\text{C}$

(c) $\Delta T_b = K_b.m_s$ = (0.51)(0.3084) = 0.157°C

(d) π at 25°C, taking R = 0.08314 L.atm/(mol.K)
π = (0.08314)(273 + 25)(0.3084) = 7.64 atm

12.3. OBJECTIVES

1. To determine the colligative properties of a nearly ideal solution as a function of concentration.

2. To determine the colligative properties of a fluid food product.

3. To estimate the relative activity of the model solution and the apparent molarity of the food product.

12.4. APPARATUS

1. A cryoscope to measure freezing point
2. Boiling point measuring apparatus
3. Glassware
4. Osmotic pressure apparatus

12.5. PROCEDURE

1. Follow the operating instructions of various equipment used in this lab.

2. Determine the freezing point of water and three different concentrations of a simple sugar. Also determine the freezing point for concentrated and two dilutions of a liquid food. Note the product label information.

3. Determine the boiling point of pure water and three concentrations of the liquid food.

4. Determine the osmotic pressure for water and three concentrations of the liquid food.

12.6. RESULTS AND DISCUSSION

1. Report the observed and predicted properties values for water, sugar solutions, and food. Use water and sugar solutions as calibrating fluids. Draw the calibration graphs by plotting observed versus calculated properties for water and sugar solutions.

2. Report the liquid food information, composition, and appearance.

3. Considering one property data, calculate the remaining colligative properties, assuming effective concentrations based on ideal behavior or the dilutest solution. Show your calculations.

4. Estimate and report the activity coefficients (γ) at various concentrations of the liquid food.

5. Calculate boiling point elevation and osmotic pressure for a 10% fruit juice (90% water) when freezing point depression is -0.65°C, K_f = 1.86°C/mol, and K_B = 0.51°C/mol.

Prelab Questions for Laboratory 12

NAME:____________________________________ **DATE:**____________________

ANSWER PRELAB QUESTIONS ON THIS SHEET

Laboratory 13

SURFACE TENSION PROPERTIES

SUMMARY

This lab is to determine the surface and interfacial tensions of liquid foods and then calculate the work of cohesion, adhesion, and spreading coefficient. It provides some background material for the application of surface tension properties in emulsion and detergency.

13.1. BACKGROUND

13.1.1. Surface and Interfacial Tensions

The molecules within the bulk of a liquid are subjected to equal forces of attraction in all directions, whereas those located at the interface experience unbalanced attractive forces resulting in a net inward pull. Surface tension or surface free energy (γ_o) is defined as the work required to increase the area of a surface isothermally and reversibly by unit amount. The same considerations apply to the interface between two immiscible liquids. Interfacial tensions (γ_I) usually lie between the individual surface tensions of the two liquids under consideration. Table 13.1 provides γ_o and γ_I values for several liquids.

Table 13.1 Surface tensions (γ_o) Against Air and Interfacial Tensions (γ_I) Against Water for Liquids at 20°C (mN/m, or mJ)

Liquid	γ_o	γ_I	Liquid	γ_o	γ_I
Water	72.75	--	Bromobenzene	35.75	--
Benzene	28.88	35.0	Toluene	28.43	--
CCl_4	26.80	45.1	Chloroform	27.14	--
n-Octanol	27.50	8.5	*n*-Butanol	--	1.6
n-Hexane	18.40	51.1	*n*-Hexanol	--	6.8
n-Octane	21.80	50.8	Isobutanol	--	2.1
Mercury	485.00	375.0	Carbon Disulphide	--	48.0
Acetic acid	27.60	--	Acetone	23.70	--
Ethanol	22.30	--	Methanol	22.60	--
n-Butyric acid	26.80	--			

From D.J.Shaw, 1980, *Introduction to Colloid and Surface Chemistry*, Butterworths, by permission of Butterworth-Heinemann Ltd., Oxford, England

The surface tension of most liquids decreases with increasing temperature in a nearly linear fashion. The Ramsay and Shields equation relates surface tension and temperature.

$$\gamma = K(T_c - T - 6)\left[\frac{M.X}{\rho}\right]^{0.667} \qquad (13.1)$$

ıere T_c is the critical temperature, T is any temperature at which γ is required, K is a constant, the molar mass of the liquid, X the degree of association of the liquid, and ρ the liquid density.

13.1.2. Surface Tension Measurement

Various methods such as capillary rise, Wilhelmy plate, ring (du-Nouy tensiometer), drop volume and drop weight, are used to measure surface and interfacial tension. The principle of ring method is discussed here, which is widely used.

The du-Nouy tensiometer measures the force needed to expand a surface. A horizontal planar ring immersed in the liquid is pulled through the surface by a force supplied by a torsion balance. The detachment force is related to the surface tension by the expression

$$\gamma = \beta.F/(4\pi R) \qquad (13.2)$$

where F is the pull on the ring, R is the mean radius of the ring and β a correction factor. The β allows for the nonvertical direction of the tension forces and for the complex shape of the liquid supported by the ring at the point of the detachment. It depends on the dimensions of the ring and the nature of the interface. Generally, β is measured by calibration.

13.1.3. Adhesion, Cohesion, and Spreading

1. The work of adhesion (W_a) between two immiscible liquids (A and B) is equal to the work required to separate unit area of the liquid-liquid interface (A/B) and form two separate liquid-air interfaces and is given by the Dupre equation:

   $$W_a = \gamma_A + \gamma_B - \gamma_{A/B} \qquad (13.4)$$

   where $\gamma_{A/B}$ is an interfacial tension.

2. The work of cohesion (W_c) for a single liquid (A) corresponds to the work required to pull apart a column of liquid of unit cross-sectional area:

   $$W_c = 2\gamma_A \qquad (13.5)$$

3. The spreading coefficient (s) is defined by:

   $$s = \gamma_{W/A} - \gamma_{O/A} - \gamma_{O/W} \qquad (13.6)$$

where $\gamma_{W/A}$, $\gamma_{O/A}$, and $\gamma_{O/W}$ are surface tensions at water-air, oil-air and oil-water interfaces, respectively. The condition for initial spreading is that s be zero or positive. For example, s for *n*-hexadecane is -9.3 mN/m (72.8 - 30.0 -52.1), hence it will not spread; s for *n*-octane is 0.2 mN/m (72.8 - 21.8 - 50.8), hence it will just spread; and s for *n*-octanol is 36.8 mN/m (72.8 - 27.5 - 8.5), hence it will spread against contamination. Table 13.2 tabulates values of spreading coefficient on water for a few liquids.

Table 13.2 Values of Initial Spreading Coefficients on Water at 20°C

Liquid	s, mN/m
Ethanol	50.4
Methanol	50.1
n-Octanol	36.8
Chloroform	13.0
Benzene	8.9
Toluene	6.8
Hexane	3.4
CCl_4	1.1
n-Octane	0.2
Carbon disulphide	-6.9
Bromoform	-9.6
Methylene iodide	-26.5

From D. Harkins 1952, *The Physical Chemistry of Surface Films*. Reproduced with permission of Van Nostrand Reinhold, New York.

13.1.4. Emulsifiers

Emulsifiers and surfactants are substances that alter the surface properties of other materials by orienting itself along the interface of the two adjacent surfaces and reducing the resistance of the two substances to combine. An emulsion is the dispersion of one substance with another where the two are immiscible. Emulsions usually contain emulsifying agents to stabilize the dispersion of the two insoluble liquids. The dispersed fluid is called the discontinuous phase and the dispersing medium is the continuous phase. Emulsions can be classified as oil in water such as milk, mayonnaise, and icecream or as water in oil emulsion such as butter, cream, and margarine.

Natural emulsifiers include lecithin, lanolin, saponins, and gums. Synthetic emulsifiers are glycerine, propylene glycol, sorbitol, and numerous other fat and fatty acids. Although an infinite number of emulsifiers are possible, mono- and diglycerides account for 75% of the emulsifier use.

13.1.5. Detergency

Detergency is the theory and practice of dirt removal from solid surfaces by surface chemical means. The work of adhesion between a dirt particle (*D*) and a solid surface (*s*) is $W_{s/D} = \gamma_{D/W} + \gamma_{s/W} - \gamma_{s/D}$. The action of the detergent is to lower $\gamma_{D/W}$ and $\gamma_{s/W}$, thus decreasing $W_{s/D}$ and increasing the ease with which the dirt particle can be detached by mechanical agitation. The surfactants that adsorb at the solid-water and dirt-water interfaces will be the best detergents.

Prelab Questions

Q1. Describe the surface and interfacial tensions of liquid foods.

Q2. Define the work of cohesion, adhesion, and spreading coefficient.

Q3. What is the effect of temperature on surface tension?

Q4. Describe briefly emulsifiers and surfactants.

Q5. What is food emulsion?

Q6. What is the concept of detergency?

13.2. OBJECTIVES

1. To determine the surface and interfacial tensions of liquids at room temperature.

2. To determine the influence of concentration of surface active agents on the surface tension of water.

13.3. APPARATUS

1. Tensiometer such as ring or du-Nouy tensiometer
2. Glassware

13.4. PROCEDURE

13.4.1. Surface Tension of Liquids

Using the ring tensiometer (calibrated), determine the dial reading at which the immersed ring pulls free from the liquid surface (distilled water, oils such as Mazola, safflower). Calculate the surface tension of the liquids. If correction factor (ß) is unknown, then calculate it by measuring the force required to pull the ring through a liquid of known surface tension (Table 13.1) and equation 13.2. Record the data in Table 13.3.

13.4.2. Effect of Surface Active Agents on the Surface Tension of Water

Using the tensiometer, determine the surface tension of solutions containing different concentrations of surface active agents.

13.4.3. Interfacial Tension of Liquids

Take 10 mL water in each of 25 mL beakers and carefully add 10 mL oil to one beaker and 10 mL benzene to the other beaker. Determine the dial reading (force) required to pull the ring through the liquid interface.

13.5. RESULTS AND DISCUSSION

13.5.1. Surface Tension of Liquids

Report the surface tension values and compare with values reported in the literature.

13.5.2. Effect of Surface Active Agents

Calculate the surface tension for each solution. Report the results of the influence of surface agents in the form of a plot of log (concentration) versus surface tension. Calculate work of cohesion for each solution.

13.5.3. Interfacial Tension

1. Calculate the interfacial tension for each interface.
2. Calculate the work of adhesion and spreading coefficient.
3. Briefly discuss why interfacial tension between two liquids and the spreading coefficient are important in forming emulsions.

13.4.2. Effect of Surface Active Agents on the Surface Tension of Water

Using the tensiometer, determine the surface tension of solutions containing different concentrations of surface active agents.

13.4.3. Interfacial Tension of Liquids

Take 10 mL water in each of 25 mL beakers and carefully add 10 mL oil to one beaker and 10 mL benzene to the other beaker. Determine the dial reading (force) required to pull the ring through the liquid interface.

13.5. RESULTS AND DISCUSSION

13.5.1. Surface Tension of Liquids

Report the surface tension values and compare with values reported in the literature.

13.5.2. Effect of Surface Active Agents

Calculate the surface tension for each solution. Report the results of the influence of surface active agents in the form of a plot of log (concentration) versus surface tension. Calculate work of cohesion for each solution.

13.5.3. Interfacial Tension

1. Calculate the interfacial tension for each interface.
2. Calculate the work of adhesion and spreading coefficient.
3. Briefly discuss why interfacial tension between two liquids and the spreading coefficient are important in forming emulsions.

Table 13.3 Data Sheet for Surface Tension Laboratory

1. Surface tension of liquids

Sample	Temperature, °C	Dial Readings 1	2	Ave.	Corrected Reading from the Calibration Curve
Water					
Mazola oil					
Safflower oil					

2. Effect of surface active agents on the surface tension of water

Surfactant ______________________

Surfactant Weight, %	Temperature, °C	Dial Readings 1	2	Ave.	Corrected Reading
0.001					
0.01					
0.1					
1.0					

3. Interfacial tension of liquids

Interface	Temperature, °C	Dial Readings 1	2	Ave.	Corrected Reading
Oil/water					
Benzene/water					

Prelab Questions for Laboratory 13

NAME:____ **DATE:**__________________

ANSWER PRELAB QUESTIONS ON THIS SHEET

Laboratory 14

THERMAL PROCESSING OF FOODS

SUMMARY

This lab is designed to provide the basis and applications of the graphical and formula methods for calculating thermal process time of foods packaged in containers. The students will collect the time-temperature data for a product at its slowest heating point and calculate the process time using both the methods.

14.1. BACKGROUND

14.1.1. Introduction

Thermal processing is a very common method of food preservation. This includes cooking, blanching, pasteurization, and sterilization. The cooking can be further classified into baking, broiling, roasting, boiling, frying (pan or oil), and stewing. Blanching is required to inactivate enzymes and is generally done before freezing. The aim of pasteurization is to destroy pathogenic vegetative cells. The objective of commercial sterilization processes, is to significantly reduce the probability for survival of microorganisms of concern to public health.

Most of the thermal processing criteria and conditions are based on the destruction of microorganisms during storage, handling, and transportation. If product safety is maintained, it can also be based on desired color, texture, nutrient retention, or flavor.

Conventional canning and aseptic processing are two commonly used thermal processes. In canning, the food is first placed in a container (can, jar, tray, etc.) and then heated whereas in aseptic processing, the product is heated to accomplish sterility and then placed in a sterile container and sealed.

To determine the time-temperature profile of a food product, the temperature changes at the slowest heating point are recorded during heating/cooling by placing a temperature sensor into the product at that location.

14.1.2. Microbial Inactivation

At a constant temperature, the rate of change in the number of viable organisms (*dN/dt*) is proportional to the number of viable organisms (*N*) present,

$$\frac{dN}{dt} = - K.N \tag{14.1}$$

where K is a reaction rate constant. Taking $N = N_o$ at $t = 0$, and $N = N$ at $t = t$, equation 14.1 can be integrated to

$$\ln\left(\frac{N}{N_o}\right) = - K.t \quad or \quad \log\left(\frac{N}{N_o}\right) = \frac{-K.t}{2.303} \tag{14.2}$$

Taking $D = 2.303/K$ = thermal death time, equation 14.2 is written as

$$\log\left(\frac{N}{N_o}\right) = \frac{-t}{D} \quad or \quad \log\left(\frac{N_o}{N}\right) = t/D \tag{14.3}$$

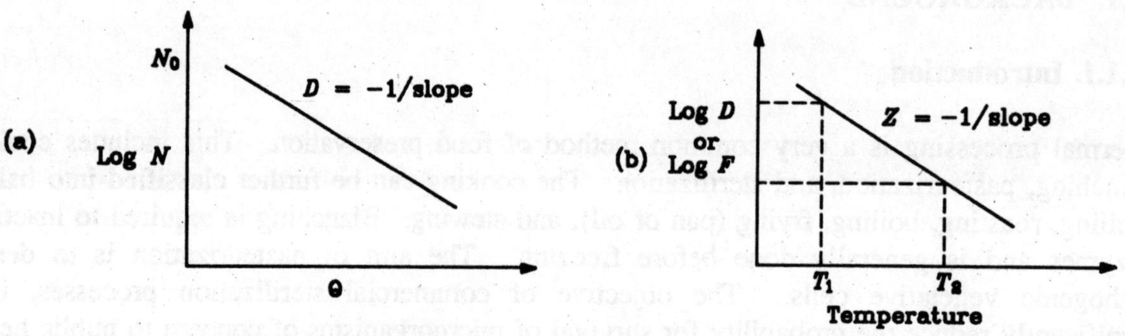

Fig. 14.1 **(a) Concept of *D* value and (b) *Z* value for microbial growth/decay.**

where D is the time required to reduce the microbial population by a factor of 10. It is the negative inverse of the slope of the plot between log N and t (i.e., D = -1/slope) (Fig. 14.1a).

Log(N/N_o) is called the sterilizing value (SV) or the number of decimal reductions. For example, a value of log (N/N_o) = 6 means that the probability of spoilage is 10^{-6}. Commercial thermal processes are designed for the inactivation of microorganisms that endanger public health and reduce microorganisms to a level that gives a very low spoilage probability.

The F value is the heating time t at temperature T to achieve a desirable sterilization. F at 250°F or 121°C is denoted by F_o. Thus,

$$F_T = (SV)\ (D_T), \text{ at temperature } T \tag{14.4}$$

D_T can be calculated from equation 14.3 after measuring N_1 and N_2 at time t_1 and t_2, that is,

$$D_T = \frac{t_2 - t_1}{\log(N_1/N_2)} \quad (14.5)$$

The rate of inactivation of microorganisms increases in a logarithmic rate with increasing temperatures. Thus, *D* or *F* values will decrease logarithmically with increase in temperature. *Z* is the change in temperature that accompanies a 10-fold change in the time for inactivation (i.e., *D* or *F* value). *Z* is calculated from the slope of the plot between log *D* or log *F* versus temperature and is given by (Fig. 14.1b)

$$Z = \frac{T_1 - T_2}{\log(F_2/F_1)} \quad (14.6)$$

or

$$\log \frac{F_2}{F_1} = \frac{T_1 - T_2}{Z}$$

Similarly,

$$\log \frac{D_2}{D_1} = \frac{T_1 - T_2}{Z} \quad (14.7)$$

14.2. PROCESS TIME

14.2.1. The General Graphical Method

This involves graphical integration of the time-temperature (heat penetration) data at the slowest heating (cold) point during thermal processing to obtain the total lethality provided to the product. The following is the basis of this method:

From equation 14.6, taking temperature in °F, we get

$$\log \frac{F_o}{F_T} = \frac{T - 250}{Z}, \text{ for a particular Z value}$$

or

$$F_o = F_T \; 10^{(T-250)/Z}$$

or

$$dt_o = dt_T \; 10_{(T-250)/Z}$$

$$F_o = \int dt_o = \int 10^{(T-250)/Z}\, dt_T$$

or

$$F_o = \Sigma \Delta t_T\, 10^{(T-250)/Z}$$

where F_o is the F value at 250°F (121°C) of a microorganism whose Z value is given, Δt_T, is the time increment at T, T is the mean temperature within a time increment, and $L = 10^{(T-250)/Z}$ = lethality function, which is the equivalent time of heating at 250°F for 1 minute of heating at temperature T. Thus, the sterilization value of a process can be computed by converting process time at any temperature to an equivalent process time at a reference temperature (generally 250°F).

To calculate process time, the heat penetration curve (temperature vs. time) at the slowest heating location of the food is required. First, subdivide the heating (cooling also if required) plot into small time increments. After calculating the arithmetic mean temperature for each time increment, the lethality for each mean temperature is computed. Each lethality value is then multiplied by the time increment and added to get F_o value. Alternatively, a plot of lethality against time is plotted. The required process time is obtained when the area under this plot equals the designed F_o.

14.2.2. The Formula Method

This is more rapid and as accurate as the graphical method. The basis of this method is to be discussed. The product temperature T_p at the slowest heating location can be written approximately by using a lumped system approach as

$$\log\left(\frac{T_r - T_p}{T_r - T_i}\right) = \frac{-\,U.A}{2.3m.c_p}\, t \qquad (14.8)$$

where

T_r = retort temperature
T_i = inital product temperature
U = overall heat transfer coefficient
A = surface area of the container in which food is placed
M = mass of the product
C_p = specific heat of the product

There will be a lag from the time of first introduction of steam to the time the heat penetrates to the center. To correct this lag, a lag factor j_h is included:

$$\log \frac{T_r - T_p}{j_h(T_r - T_i)} = \frac{-U.A}{2.3m.C_p} t \qquad (14.9)$$

The plot of $\log(T_r - T_p)$ vs. t will be linear (Fig. 14.2) with an intercept. Taking f_h = -1/slope = 2.3 $(m.C_p)/(U.A)$ = time of heating required for the temperature-time plot to traverse one log cycle, equation 14.9 can be written as

$$\log \frac{T_r - T_p}{j_h(T_r - T_i)} = - \frac{t}{f_h} \qquad (14.10)$$

where

$$j_h = \frac{(T_r - T_p) at\ t = 0.6\ of\ retort\ come-up\ time}{T_r - T_o} \qquad (14.11)$$

The retort come-up time is the time lag from the time steam is turned on to the time retort reaches process temperature. Only 40% of the retort come-up time has any heating value. T_o is the apparent initial temperature of the product (Fig. 14.2).

The value of g is the maximum temperature difference, $(T_r - T_p)_{max}$, at the end of the heating cycle after thermal process time (t). Mathematically, it is written as

$$\log(T_r - T_p) = \frac{-t}{f_h} + \log\ [j_h(T_r - T_i)]$$

or

$$\log(g) = \frac{-t}{f_h} + \log\ [j_h(T_r - T_i)]$$

or

$$t = f_h \log[j_r(T_r - T_i)/g] \qquad (14.12)$$

This is the basis of the formula method. The cooling curve can also be analyzed similarly to that used for heating. This will provide f_c and j_c values.

To calculate f_h and j_h, the product temperature can be directly plotted on the right-hand axis, since the left-hand axis is $T_r - T_p$ (Fig. 14.2). The U is the desired process time at T_r and is given by

$$U = F_o\ 10^{(250-Tr)/Z} \qquad (14.13)$$

The g depends on f_h, Z, process time, and $T_r - T_c$, where T_c is cold water temperature. To include cooling in the calculations, Stumbo (1973) provided tables to calculate g as a function of j_c, Z,

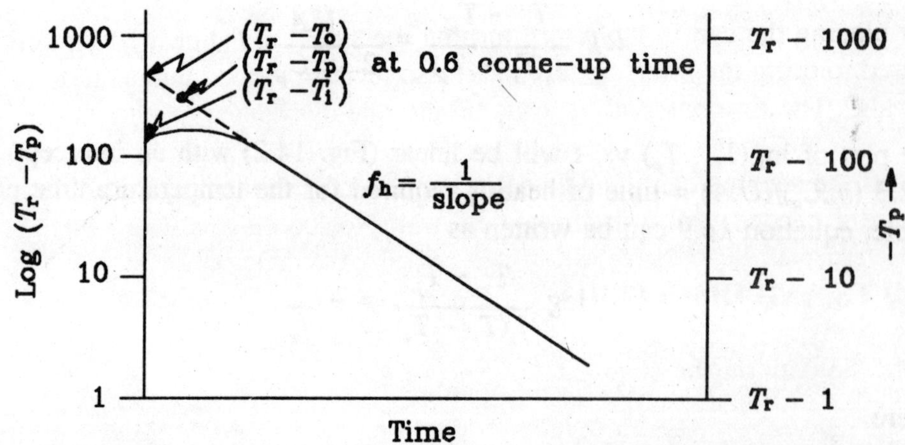

Fig. 14.2 Computation of f_h and j_h values.

and f_h/U at $I_c + g$ = 180°F, where $I_c = T_p - T_c$. If a correction is required for other $I_c + g$ values, the actual F_o can be calculated by subtracting 1% from the theoretical F_o for every 10°F the $I_c + g$ is above 180°F, and by adding 1% for every 10°F the $I_c + g$ is below 180°F. Table 14.1 provides g for various f_h/U and j_c for Z = 18°F. For other Z values and for calculating process time for foods having a broken heating curve, see Stumbo (1973).

14.3. APPARATUS

Two agitated control temperature water baths at T_r and T_c or a retort with appropriate instrumentation and controls, vernier caliper, spatula, weighing balance, can opener, can sealing machine, cans, food products, Ecklund thermocouple with fittings, and temperature recording system.

14.4. PROCEDURE

1. Fill the can 90% full with the food product after installing an Ecklund thermocouple assembly from the side of the can at its geometric center. Be sure the gaskets are tightened. Seal the can properly.

2. Measure and record dimensions and mass of the can. Connect the thermocouple leads to the recorder.

3. Place the can in the retort. If retort is not available, a hot water bath at a constant temperature (85 to 95°C) can be used for heating cycle and a cold temperature bath for cooling cycle.

4. After placing the can in the retort, turn on the steam at the desired T_r. Record the time required to bring the retort up to T_r, and also record the food temperature (T_p) at frequent intervals. The time interval depends on the rate of product heating.

5. Turn off the steam when the cold point temperature is within 5°C of T_r. Maintain the air pressure if needed, and turn on the cold water for cooling. Record the T_p up to 40°C.

14.5. RESULTS AND DISCUSSION

1. Record the data on the attached data sheet (Table 14.2)

2. Calculate f_h, j_h, f_c, and j_c after plotting T_p vs. t as discussed in Section 14.2.2.

3. Calculate the thermal process time at a given T_r, Z = 18°F, F_o = 8 min using the formula method, and with and without the cooling effect.

4. Repeat step 3 by using the graphical method as given in Section 14.2.1. Compare the results with step 3.

5. Using the computer programs "Thermal Processing", calculate the process time using both methods and compare the results obtained in steps 3 and 4.

Prelab Questions

Q1. Define 'D', 'Z' and 'SV'.

Q2. What is the basis of the graphical method?

Q3. What is a lethality function?

Q4. What is a retort come-up time?

Q5. What is the basis of the Formula method?

Table 14.1 Values of g for Various j_c and f_h/U Values at Z = 18°F

	Values of g When j of Cooling Curve Is:								
f_h/U	0.40	0.60	0.80	1.00	1.20	1.40	1.60	1.80	2.00
0.20	4.09–5[a]	4.42–05	4.76–05	5.09–05	5.43–05	5.76–05	6.10–05	6.44–05	6.77–05
0.30	2.01–03	2.14–03	2.27–03	2.40–03	2.53–03	2.66–03	2.79–03	2.93–03	3.06–03
0.40	1.33–02	1.43–02	1.52–02	1.62–02	1.71–02	1.80–02	1.90–02	1.99–02	2.09–02
0.50	4.11–02	4.42–02	4.74–02	5.06–02	5.38–02	5.70–02	6.02–02	6.34–02	6.65–02
0.60	8.70–02	9.43–02	1.02–01	1.09–01	1.16–01	1.23–01	1.31–01	1.38–01	1.45–01
0.70	0.150	0.163	0.176	0.189	0.202	0.215	0.228	0.241	0.255
0.80	0.226	0.246	0.267	0.287	0.308	0.328	0.349	0.369	0.390
0.90	0.313	0.342	0.371	0.400	0.429	0.458	0.487	0.516	0.545
1.00	0.408	0.447	0.485	0.523	0.561	0.600	0.638	0.676	0.715
2.00	1.53	1.66	1.80	1.93	2.07	2.21	2.34	2.48	2.61
3.00	2.63	2.84	3.05	3.26	3.47	3.68	3.89	4.10	4.31
4.00	3.61	3.87	4.14	4.41	4.68	4.94	5.21	5.48	5.75
5.00	4.44	4.76	5.08	5.40	5.71	6.03	6.35	6.67	6.99
6.00	5.15	5.52	5.88	6.25	6.61	6.98	7.34	7.71	8.07
7.00	5.77	6.18	6.59	7.00	7.41	7.82	8.23	8.64	9.05
8.00	6.29	6.75	7.20	7.66	8.11	8.56	9.02	9.47	9.93
9.00	6.76	7.26	7.75	8.25	8.74	9.24	9.74	10.23	10.73
10.00	7.17	7.71	8.24	8.78	9.32	9.86	10.39	10.93	11.47

Table 14.1 (Continued) Values of '*g*' for Various *j* and f_h/U Values at *Z* = 18°F

f_h/U	Values of *g* When *j* of Cooling Curve Is:								
	0 40	0.60	0.80	1.00	1.20	1.40	1.60	1.80	2.00
15.00	8.73	9.44	10.16	10.88	11.59	12.31	13.02	13.74	14.45
20.00	9.83	10.69	11.55	12.40	13.26	14.11	14.97	15.82	16.68
25.00	10.7	11.7	12.7	13.6	14.6	15.6	16.5	17.5	18.4
30.00	11.5	12.5	13.6	14.6	15.7	16.8	17.8	18.9	19.9
35.00	12.1	13.3	14.4	15.5	16.7	17.8	18.9	20.0	21.2
40.00	12.8	13.9	15.1	16.3	17.5	18.7	19.9	21.1	22.3
45.00	13.3	14.6	15.8	17.0	18.3	19.5	20.8	22.0	23.2
50.00	13.8	15.1	16.4	17.7	19.0	20.3	21.6	22.8	24.1
60.00	14.8	16.1	17.5	18.9	20.2	21.6	22.9	24.3	25.7
70.00	15.6	17.0	18.4	19.9	21.3	22.7	24.1	25.6	27.0
80.00	16.3	17.8	19.3	20.8	22.2	23.7	25.2	26.7	28.1
90.00	17.0	18.5	20.1	21.6	23.1	24.6	26.1	27.6	29.1
100.00	17.6	19.2	20.8	22.3	23.9	25.4	27.0	28.5	30.1
150.00	20.1	21.8	23.5	25.2	26.8	28.5	30.2	31.9	33.6
200.00	21.7	23.5	25.3	27.1	28.9	30.7	32.5	34.3	36.2
250.00	22.9	24.8	26.7	28.6	30.5	32.4	34.3	36.2	38.1
300.00	23.8	25.8	27.8	29.8	31.8	33.7	35.7	37.7	39.7
350.00	24.5	26.6	28.6	30.7	32.8	34.9	37.0	39.0	41.1
400.00	25.1	27.2	29.4	31.5	33.7	35.9	38.0	40.2	42.3

Table 14.1 (Continued) Values of 'g' for Various j and f_h/U Values at Z = 18°F.

f_h/U	Values of g When j of Cooling Curve Is:								
	0.40	0.60	0.80	1.00	1.20	1.40	1.60	1.80	2.00
450.00	25.6	27.8	30.0	32.3	34.5	36.7	38.9	41.2	43.4
500.00	26.0	28.3	30.6	32.9	35.2	37.5	39.8	42.1	44.4
600.00	26.8	29.2	31.6	34.0	36.4	38.8	41.2	43.6	46.0
700.00	27.5	30.0	32.5	35.0	37.5	39.9	42.4	44.9	47.4
800.00	28.1	30.7	33.3	35.8	38.4	40.9	43.5	46.0	48.6
900.00	28.7	31.3	34.0	36.6	39.2	41.8	44.4	47.0	49.7
999.99	29.3	31.9	34.6	37.3	39.9	42.6	45.3	47.9	50.6

[a] 4.09–05 means 4.09 x 10^{-5}

From C.R. Stumbo 1973, *Thermobacteriology in Food Processing*. Reprinted by permission of Academic Press, Orlando, FL.

Table 14.2 Data Sheet for Temperature History

Can dimensions: length ____________________ cm, diameter _______________ cm

Can mass: filled __________________ g, empty ________________ g

Food mass in the can: __________________ g

Food product composition: ____

Other food product information such as pH, water activity, apparent viscosity or consistency, etc.:

____

____

Retort temperature (T_r) ________________ °C, Initial product temperature ____________°C

Retort come-up-time ______________________min

No.	Time,min	Product Temperature (T_p), °C
1		
2		
3		
4		
5		
6		
7		
8		
9		
10		
11		

(continued)

Table 14.2 (continued)

No.	Time, min	Product Temperature (T_p), °C
12		
13		
14		
15		
16		
17		
18		
19		
20		

Prelab Questions for Laboratory 14

NAME:______________________________ DATE:____________________

ANSWER PRELAB QUESTIONS ON THIS SHEET

Laboratory 15

DRYING CHARACTERISTICS OF FOODS

SUMMARY

In this lab you will study the dehydration behavior of a food product. By graphical analysis, you will determine various drying rate periods for the product.

15.1. BACKGROUND

15.1.1. Drying

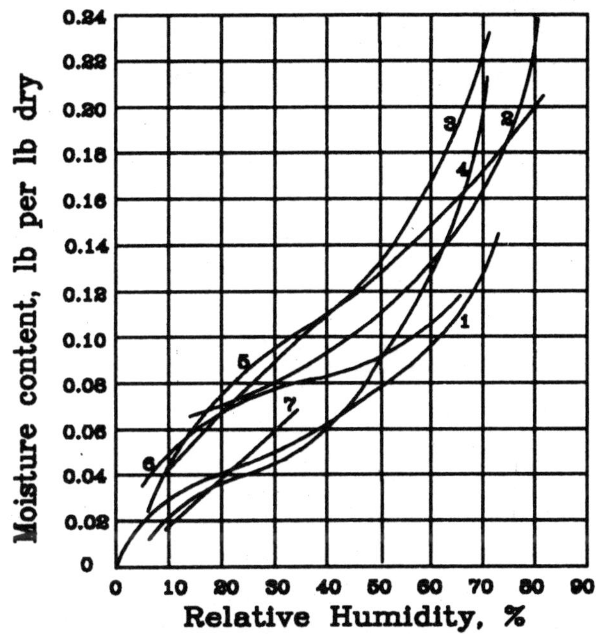

Fig. 15.1 Selected water sorption isotherms. 1–Egg solids, 10°C (from Gane, 1943); 2–beef, 10°C (from Bate–Smith *et al.*, 1943); 3–fish (cod), 30°C (from Jason, 1958), 4–coffee, 10°C (from Gane, 1950); 5–starch gel, 25°C (from Fish, 1958); 6–potato, 28°C (from Gane, 1950); 7–orange juice (from Notter *et al.*, 1959). From W.B. Van Arsdel, M.J. Copley, and A.I. Morgan, 1973, *Food Dehydration*, Vol. 1. Reprinted by permission of Van Nostrand Reinhold, New York, N. Y.

Dehydration or drying of foods is a complex phenomenon involving momentum, heat and mass transfer, physical properties of the food, air and water vapor mixtures, and macro and microstructure of the food. There are many possible drying mechanisms, but those that control the drying of a particle product depend on its structure and the drying parameters--drying conditions, moisture content, dimensions, surface transfer rates, and equilibrium moisture content. These mechanisms fall into three classes: (1) evaporation from a free surface, (2) flow as a liquid in capillaries, and (3) diffusion as a liquid or a vapor. The first mechanism follows the laws for heat and mass transfer for a moist object. The second mechanism becomes difficult to distinguish from diffusion when one sets the surface tension potential to be proportional to the logarithm of the moisture potential (or water activity). The third set of mechanisms follows Fick's second law of diffusion, which is analogous to Fourier's law of heat transfer when the appropriate driving force is used.

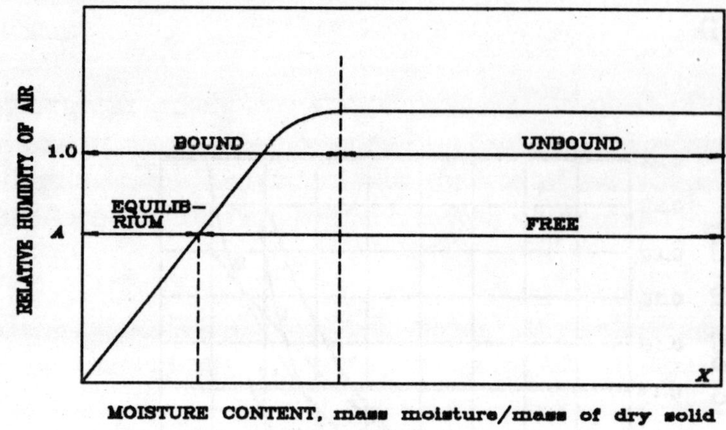

Fig. 15.2 Equilibrium moisture isotherm for a material showing various moisture contents.

All solid materials have a certain equilibrium moisture content when in contact with air at a particular temperature and humidity. The material will tend to lose or gain moisture over a period of time to attain this equilibrium value. Figure 15.1 shows typical equilibrium moisture isotherms for a few food products. The equilibrium moisture curve depends on the environment temperature for a particular material. Figure 15.2 shows various moisture content regions of a food material such as bound, unbound, and free.

In convectional drying the heating medium, generally air, comes into direct contact with the solid. Various oven, rotary, fluidized bed, spray, and flash dryers are typical examples. In conduction drying, the heating medium is separated from the solid by a hot conducting surface. Examples are drum, cone, and through dryers. In radiation dryers, the heat is transmitted as radiant energy. Some dryers also use microwave energy to dry food materials at atmospheric pressure or at vacuum.

15.1.2. Selection and Design of a Dryer

The factors required to be considered in the selection and design of a dryer are (1) material properties--thermophysical properties, transfer coefficients, flammability, and toxicity; (2) drying characteristics--initial, final, and equilibrium moisture contents, required drying time and limitations on drying temperature; (3) dried product specification--purity, physical form, and constraints regarding the physical, chemical, and surface properties. Table 15.1 provides recommended conditions for the dehydration of fruits and vegetables.

15.1.3. Rate and Time of Drying

A generalized rate of drying curve is shown in Fig. 15.3, which provides various drying rate periods. Generally, a drying curve can be divided into three drying rate periods:

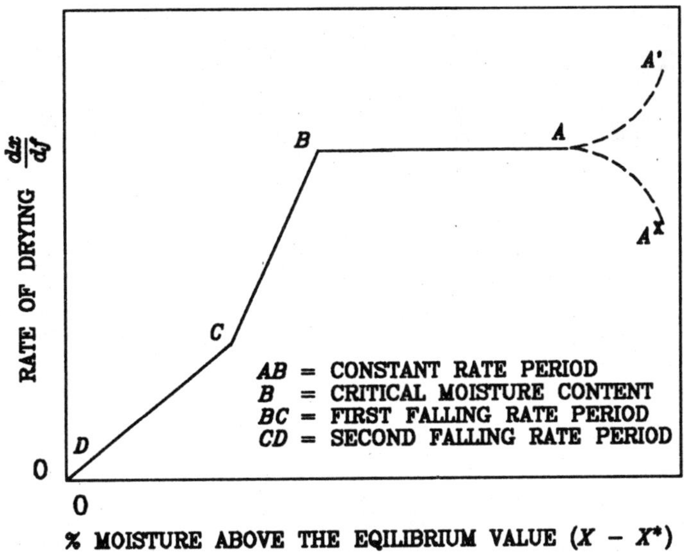

Fig. 15.3 Generalized rate of drying curve for a solid material.

1. *Constant rate period*: During this period (section *AB*) drying takes place by evaporation of moisture from a saturated surface. The following equations defines rate of evaporation (N_c) and heat transfer rate:

$$N_c = \frac{dx}{dt} = K_g.A.(p_s - p_w) = h.A.\ (T_e - T_s)\ /\ h_{fg} \tag{15.1}$$

where K_g is the overall mass transfer coefficient for the gas film, p_s is the vapor pressure at the surface temperature (T_s = wet bulb temperature of air), p_w is the vapor pressure in the air, A is the surface area, t the time, x is the moisture content in the solid, d.b., h is the heat transfer coefficient, h_{fg} is the enthalpy of evaporation, and T_e is the air temperature. In this period, the drying rate is determined by h, A, and T_e and is not affected by conditions within the solid.

Table 15.1 Recommended Conditions for Dehydration of Fruits and Vegetables

Product	Blanching Conditions	Sulfiting Conditions, ppm	kg/m^2 of Tray Area	Drier Temp, °C	Final Moisture dry basis
Asparagus	4-5 min in steam	No sulfite	7.3	57	0.053
Beans, green	2-4 min, 88°C	1000	7.3	91; 30 min 77; 30 min 66; 30 min 60; 4.5 h	0.053
Beans, lima	5-6 min in steam	No sulfite	4.9-6.1	71-78	0.053
Beets	Dice, 5-7 min at 93°C-99°C	No sulfite	19.6	1st drier: 93-99 2nd drier: 52	0.053
Broccoli	10-12 min in steam	No sulfite	4.9	71	0.031
Brussels sprouts	5 min in steam	No sulfite	4.9-6.1	71	0.031
Cabbage	93°C for about 3 min	1500-2500	4.9-6.1	82--first stage parallel flow : 63--second stage countercurrent	0.042
Carrots	Dice, steam at 88°C for 6-8 min	Finished product should contain 500-1000	6.1	Countercurrent 71	0.042
Celery	Pieces (1.3-1.9 cm) steam 1-2 min	No sulfite	4.9-6.1	Countercurrent 52-57	0.042
Sweet corn	99°C for about 2 min or until peroxidase is inactivated	Dry product contains 2000	7.8	Parallel flow, hot end (82) cool end (74) with wet bulb (43); counter-current, hot end (74 dry bulb, wet bulb 35)	0.053

(continued)

Table 15.1 (Continued)

Spinach, kale, chard and mustard and turnip greens	Steam for about 2 min	No sulfite	3.7–6.5	Parallel flow tunnel (93) hot end, (82) cold end and a wet bulb of 49; counter-current hot end (77) wet bulb (38)	0.042
Peas	Boiling water 1–2 min	No sulfite	4.9	Parallel flow, hot end (82) cold end (71) and wet bulb (43) at cool end; countercurrent dry bulb at hot end (66) and wet bulb at cool end (38)	0.053
Potato	Steam (93°C–99°C) for 3–6 min or until peroxidase inactivated	0.2–1.0% Solution of a sulfite salt	7.3	71 hot end	0.075
Apple slices or rings	No blanch	Dip in 0.5% bisulfate solution	7.3–9.8	1st stage; 74 2nd stage; 71	0.177–0.316
Apricots	No blanch	Finished product 2000 ppm, but not more than 2500 ppm	9.8	66	0.177–0.250
Peaches	No blanch	1500 ppm absorbed	12.2–14.7	Countercurrent, 68 with 8° difference between dry and wet bulb	0.250
Pears	No blanch or blanch	1500 ppm absorbed	14.7	66	0.177–0.35
Figs	No blanch	No sulfite	9.8–14.7	60–66	0.177–0.250
Plums (prunes)	Dipping	No sulfite	14.7	French (66–74) Imperial (60)	0.136–0.234

From M.L. Fields 1977, *Laboratory Manual in Food Preservation*. Reprinted with permission of Van Nostrand Reinhold Publ. Co., New York.

2. *First falling rate period*: Point B, the moisture content at the end of the constant rate period, is the "critical moisture content." At this point the surface of the solid is no longer saturated, and the rate of drying decreases with the decrease in moisture content. At point C, the surface moisture film has evaporated fully, and with the further decrease in moisture content, the drying rate is controlled by the rate of moisture movement through the solid.

3. *Second falling rate period*: Period C to D represents conditions when the drying rate is largely independent of conditions outside the solid. The moisture transfer may be by any combination of liquid diffusion, capillary movement, and vapor diffusion.

The drying time in different stages of drying can be calculated by the following procedure:

If W_o = mass of dry solid, then the moisture transfer rate (N_A) is

$$N_A = -W_o \frac{dx}{dt} \tag{15.2}$$

For constant drying rate period (A to B); the drying time (t_{AB}) is:

$$t_{AB} = W_o(X_A - X_B)/N_c \tag{15.3}$$

For the first falling rate period (B to C), the drying time (t_{BC}) can be calculated if moisture transfer rate (N) is represented by $N = m.X + k$, where m is the slope of the line and k is its intercept, where $N_B = m.X_B + k$ and $N_c = m.X_c + k$:

$$t_{BC} = W_o \frac{x_B - x_C}{N_B - N_C} \ln\left(\frac{N_B}{N_C}\right) \tag{15.4}$$

Similarly, drying time for the second falling rate period can be calculated.

15.1.4. Moisture Contents on Wet and Dry Basis

The moisture content of a product can be represented on the basis of the wet or dry mass of the product:

$$\text{Moisture content wet basis } (X_{wb}) = \frac{\text{mass of moisture}}{\text{mass of wet product or initial mass of the product}} \tag{15.5}$$

and

$$\text{Moisture content dry basis } (X_{db}) = \frac{\text{mass of moisture}}{\text{(mass of wet product-mass of moisture) or mass of dry matter}} \tag{15.6}$$

also

$$X_{db} = X_{wb} / (1 - X_{wb}) \tag{15.7}$$

Prelab Questions

Q1. A vegetable contains 80% moisture on a wet basis. How much is the % of dry matter in the vegetable?

Q2. What are some problems of quality that may occur when foods are dried?

Q3. What are mass transfer properties important in food processing?

Q4. What is the influence of shrink on the movement of moisture in the food product?

Q5. What factors influence the mass transfer coefficient for a process system?

15.2. OBJECTIVES

1. To determine the moisture and moisture loss rate histories of food products in a forced air dryer.

2. To analyze the moisture loss rate as a function of moisture content X to determine whether (a) the constant drying rate period exists and (b) the falling rate period can be broken up into two or more regions with apparently different moisture transfer mechanisms.

3. To determine whether any portions of the moisture histories can be linearized by plotting log $(X - X_e)$ vs. time.

15.3. APPARATUS

1. Forced air or other types of dryer
2. Weighing balance
3. Temperature recording system
4. Anemometer or other air flow metering device

15.4. PROCEDURE

1. Measure the critical dimensions of the samples and calculate the surface area and volume. Weigh the sample to be dried.

2. Determine the moisture content of the sample in a vacuum oven.

3. Switch on the air blower and adjust to the required flow rate by means of the damper or other means.

4. Switch on the air heaters and adjust to the required temperature. Allow 10 min for the apparatus to attain steady-state conditions.

5. Measure and record dry and wet bulb temperature of the heated air and air flow rate.

6. Start the stopwatch and record the sample mass at time zero. Measure the sample mass at short time intervals initially (1 min), gradually extending the period as drying progresses (5 min) intervals after about 15 min. Continue the drying process for at least 45 min.

7. Take the final mass of the sample after drying for 24 h to calculate equilibrium moisture content. Record the data in Table 15.2.

15.5. RESULTS AND DISCUSSION

1. Plot the sample mass versus time and determine the equilibrium mass by extrapolation.

2. Determine the mass of solids in the sample dried in the dryer based on the initial moisture content, X_o. Also calculate the equilibrium moisture content, X_e. Calculate the X, $X - X_e$, and $\Delta X/\Delta t$ for each time in the table. $\Delta X/\Delta t$ = moisture loss $(\Delta m)/(\Delta t.\text{mass of solids})$.

3. Plot the rate of moisture loss $\Delta X/\Delta t$ versus time and show various drying rate periods.

4. Plot $\Delta X/\Delta t$ vs. $(X - X_e)$ and show various drying rate periods.

5. Plot $(X - X_e)$ versus time on a semilog coordinates. Calculate j and f.

$$j = \frac{(X - X_e)\text{ apparent}}{(X - X_e)\text{ original}}, \quad f = \text{time for one log cycle change in } X - X_e$$

6. Calculate the apparent moisture diffusivity (D_m) of the sample using

$$f = 2.303R^2\beta^2/(\pi^2 D_m)$$

where

R is the radius of an equivalent sphere of equal volume $(V = 4/(3\ \pi R^3))$

β is shape factor, 1.0 for sphere

Table 15.2 Data Sheet for the Drying Characteristics of Foods

1. Sample moisture content

   Initial mass ______________________ g; Dried mass ____________________ g

   Moisture content, wet basis ______________ %; dry basis ______________ %
   Dry matter mass in the dryer ______________ g

2. Sample dimensions ___________ m ___________ m ___________ m

   Surface area __________________ m^2; Volume ____________________ m^3

3. Air conditions

   Dry bulb temperature _____________ °C; Wet bulb temperature _____________ °C
   Air flow rate ______________

4. Equilibrium moisture content, dry basis (X_e) ____________________

Time, min	Sample Mass, g	Mass of Moisture, g	Moisture Loss, g	Moisture Loss Rate, $\Delta X/\Delta t$	Moisture Content, X	Free Moisture, $X-X_e$

Prelab Questions for Laboratory 15

NAME:__________________________________ DATE:__________________________

ANSWER PRELAB QUESTIONS ON THIS SHEET

Laboratory 16

PERFORMANCE OF A REFRIGERATION SYSTEM

SUMMARY

This laboratory exercise describes the operation, principles, and performance characteristics of a refrigeration system. This will provide practice to use refrigerant properties tables and plotting the cycle on p-h and T-S diagrams.

16.1. BACKGROUND

A refrigeration system absorbs heat at a given temperature and discharges it at a higher temperature. Mechanical energy or its equivalent is required to accomplish this. Two types of refrigeration systems are commonly used in present day perishable product storage. The most common type is a direct expansion system shown schematically in Fig. 16.1. Flooded systems are also used for cold storage of perishables, but their use is restricted to relatively few large storage facilities.

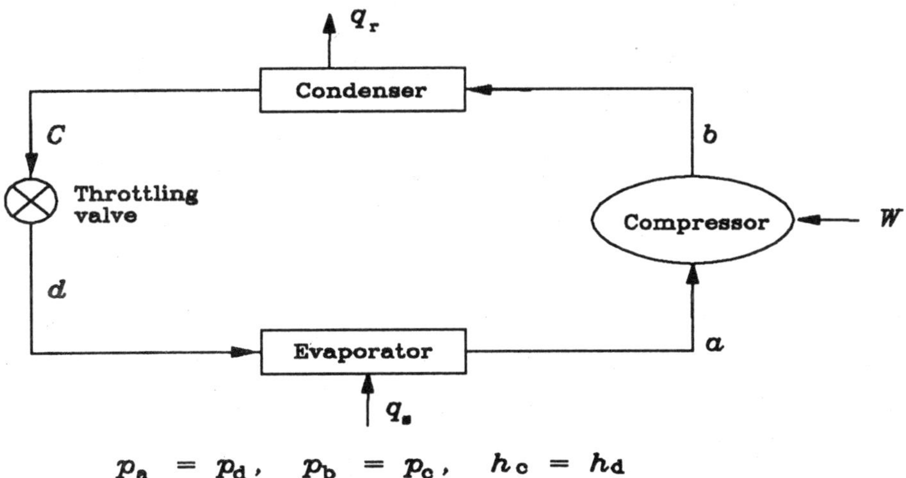

Fig. 16.1 Components of a vapor compression refrigeration.

In a typical vapor compression refrigeration system, high pressure liquid refrigerant flows through the expansion valve and boils under reduced pressure within the evaporator. Heat is absorbed by the evaporating refrigerant, resulting in the cooling of the space surrounding the evaporator coil. The evaporation temperature is determined by the internal evaporator pressure and the type of refrigerant used. The rate of heat removal depends on the refrigerant flow rate, the internal coil pressure, and the surface area of the evaporator coil. An expansion valve divides the high and low pressure sides of the direct expansion system. In commercial systems a thermostatic expansion valve controller senses the temperature of the suction line and modulates the expansion valve opening. The thermostatic expansion valve is sized to match the evaporator capacity and prevent liquid refrigerant from returning to the compressor.

An evaporator pressure regulator controls the refrigerant pressure and temperature inside the evaporator. This device is set to maintain the desired evaporator coil temperature during long- term storage. The cold, low pressure vapor from the accumulator enters the compressor where the pressure and temperature are increased. The hot, compressed vapor is then cooled in the condenser and the refrigerant returns to a liquid state. The heat gained during the compression process plus the heat absorbed during the evaporation process is rejected during condensation. The components of the refrigeration system can be divided into three categories: (1) necessary components for system operation (i.e., compressor, condenser, expansion valve, and evaporator), (2) generally used accessory equipment (i.e., storage tank, filter/dryer, slight glass, etc.), and (3) equipment used for control and monitoring of system--watt-meter, pressure gauges, flow meter, temperature recorder. The watt-meter is used to indicate the amount of power used by the compressor. Pressure gauges indicate the high and low side pressures for the system. Most refrigeration systems are designed so that the pressure loss through the pipes and heat exchangers are negligible. The two points where significant pressure changes occur are at the expansion valve and across the compressor. The temperature of the saturated portion of the evaporator and condenser can be read from the refrigerant properties tables against known pressure values. All other temperatures must be determined by direct measurement. Temperatures of the cold gas entering the compressor and hot gas leaving the compressor, of the liquid before entering the expansion valve, and of the water or air entering and leaving the evaporator should be recorded. The flow meters indicate the liquid refrigerant flow rate and the air or water flow rate through the evaporator.

Figure 16.2 shows the vapor compression refrigeration cycle on *T-S* and *h-p* diagrams. Saturated or superheated vapor enters the compressor at *a* and leaves as superheated vapor at *b*. Heat rejection and addition occur at constant pressure. The throttling process (*c-d*) produces no work. Thus, the enthalpy of the regrigerant remains constant ($h_c = h_d$). The vapor may be super-cooled prior to the expansion.

The refrigeration effect or heat addition (q_s) is

$$q_s = h_a - h_d = h_a - h_c \tag{16.1}$$

and heat rejection (q_r) is

$$q_r = h_c - h_b \tag{16.2}$$

Similarly, the work *w* of isoentropic (constant entropy) process is

$$w = h_a - h_b \tag{16.3}$$

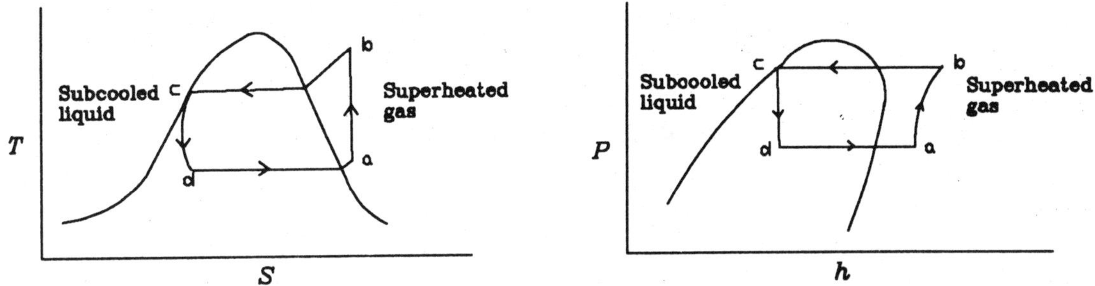

Fig. 16.2 Vapor compression refrigeration cycle on T–S and P–h diagrams.

The coefficient of performance (COP) of the system as refrigeration (R) or heat pump (HP) is

$$\mathrm{COP_R} = q_s/|w| = \frac{h_a - h_c}{h_b - h_a} \tag{16.4}$$

$$\mathrm{COP_{HP}} = q_r/w = \frac{h_c - h_b}{h_a - h_b} \tag{16.5}$$

The rate at which heat is absorbed by a system (kW) is expressed in "tons of refrigeration." A ton of refrigeration is defined as the rate of heat absorption required to change 1 ton (2000 lb) of ice at 0°C to 1 ton of water at 0°C in one day. This is equivalent to 200 BTU/min or 3.5168 kW. The refrigerant flow rate ($\dot{m}$) required for a load of N tons is, then, the total heat absorption rate divided by the refrigerating effect per kg:

$$\dot{m} = \frac{3.5168N}{(h_a - h_c)} \tag{16.6}$$

Prelab Questions

Q1. List some common refrigerants.

Q2. Why do we want to refrigerate foods?

Q3. Draw a diagram of a refrigeration system.

Q4. Why do vegetables need to be blanched before freezing?

Q5. Mark the following by true or false:

(a) Freezing kills all the spoilage bacteria.
(b) Oxidative rancidity can occur in frozen foods.
(c) Meat respires during cold storage.
(d) A standard ton of refrigeration is 200 BTU/min.
(f) Vitamin C may be lost from foods even though the food is refrigerated.
(g) A 60% syrup would freeze faster than water.

16.2. OBJECTIVES

1. To determine the performance characteristics of a vapor compression refrigeration system.

2. To familiarize yourself with various components of a vapor compression refrigeration system.

16.3. APPARATUS

1. A demonstration unit of a refrigeration system with flow meter, pressure gauges, and temperature probes.

2. Data recording unit.

16.4. PROCEDURE

1. Adjust the expansion valve at the desired flow rate of the refrigerant. Allow 15 min after starting compressor before collecting data.

2. Maintain as nearly constant room temperature as possible.

3. Observations (P_a, P_b, T_a, T_b, and T_c, watt-meter reading) should be recorded at 5 min intervals in Table 16.1, finally selecting a representative run for calculations.

16.5. RESULTS AND DISCUSSION

1. Draw schematic of the system. Make note of the function of each component and instrument in the system.

2. Draw the refrigeration cycle on *p-h* and *T-s* diagrams.

3. Report q_r, q_s, w, COP_R and COP_{HP}, after calculating various enthalpy values from the suitable properties tables of the refrigerant used. These tables are available in Appendix C. The computer program "Table" can also be used to calculate these properties.

4. Solve the following problem.

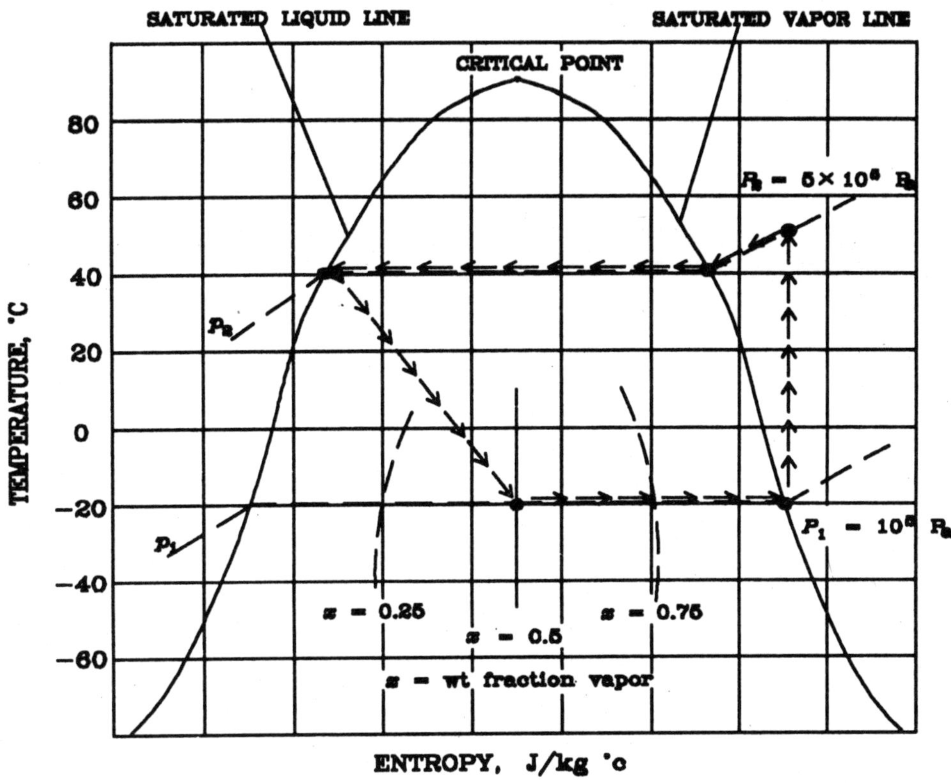

Fig. 16.3 **Special water freezer with refrigerant A.**

The attached temperature-entropy diagram (Fig. 16.3) describes the operation of a water-freezing steel unit using refrigerant A. The compressor circulates 100 kg of refrigerant A per hour. The condenser, used to cool the superheated vapor and condense the saturated vapor, is cooled with fans circulating ambient air at a temperature of 25°C. The water to be frozen is fed to the chilling plate (i.e., evaporator) (in good contact with the evaporating refrigerant) at 0°C.

Calculate the following: (state any assumptions you make)

(a) The required heat transfer area of the condenser
(b) The amount of water that can be frozen per hour
(c) The rating of the system, in tons of refrigeration

SUPPLEMENTARY INFORMATION

(a) Heat of fusion of water = 335 kJ/kg

(b) Specific heat of refrigerant A vapor = 1.05 kJ/(kg.K)

(c) Heat of vaporization of refrigerant A:

1256 kJ/kg at $p_1 = 1 \times 10^5$ Pa

1047 kJ/kg at $p_2 = 5 \times 10^5$ Pa

(d) Heat transfer coefficients in kW/(m^2.K):

For condensing or evaporating liquids	$h = 1.163$
For flowing liquids	$h = 0.582$
For flowing air	$h = 0.116$

(e) Thermal conductivity of steel = 46.5 W/(m.K)

(f) Thickness of condenser wall = 5 mm

Table 16.1 Data Sheet for a Refrigeration System Performance

Refrigerant ______________________ ________

Ambient temperature ______________________ °C

Time, min	P_a, Pa	P_b, Pa	T_a,°C	T_b,°C	T_c,°C	$\dot{m}$	Watt-meter Reading, W
5							
10							
15							
20							
25							
30							

Using the property tables or computer program for the refrigerant, calculate:

h_a ____________ kJ/kg; H_a __________ kW; s_a ____________

h_b ____________ kJ/kg; H_b __________ kW; s_b ____________

$h_c = h_d$ ____________ kJ/kg; H_c __________ kW; s_c ____________

$q_r = h_c - h_b =$ $Q_r = H_c - H_b =$

$q_s = h_a - h_c =$ $Q_s = H_a - H_c =$

Ideal work $= h_a - h_b =$ $W = H_a - H_b =$

$COP_R =$

COP_{HP}

Prelab Questions for Laboratory 16

NAME:________________________________ DATE:____________________________

ANSWER PRELAB QUESTIONS ON THIS SHEET

Laboratory 17

WATER VAPOR TRANSMISSION OF FOOD PACKAGING FILMS

SUMMARY

This lab is to determine the water vapor transmission characteristics of food packaging films. Desiccant and water methods for the measurements of water vapor permeability of packaging films are described.

17.1. BACKGROUND

The important vinyl polymer films for food packaging include Saran (copolymer of vinyl chloride and vinylidene chloride) and polyvinyl chloride. Polyfluorocarbons are costly and have unique characteristics. Fluorinated hydrocarbons have high permeability to oxygen and good inertness. On the other hand, trifluoro-chloroethylene has a low permeability to many gases and vapors. Polyvinyl fluoride has medium permeability to various gases. Water-soluble films are polyvinyl alcohol, collagen, and some cellulose derivatives and polysaccharides. Nylon films are inert and heat resistant. Packaging films controls light and oxygen, CO_2, and moisture concentrations.

Oxidation of fats and oils in food creats oxidative rancidity. Other ingredients such as some amino acids, pigments, proteins, and vitamins are also oxygen sensitive. The rate of respiration in fruits and vegetables can be reduced during storage by reducing the partial pressure of oxygen in the packaging. However, if anaerobic respiration continues for some time the fruits and vegetables spoils.

17.1.1. Permeability to Water Vapor and Gases

Water vapor and gases, such as O_2, N_2, CO_2, permeate through the packaging materials by microscopic pores or by activated diffusion due to concentration gradient. The gas and vapor permeability can be calculated by using Fick's first law: $J = -D_g.A.dc/dx$, for one- dimensional diffusion where J = rate of diffusion, mol/s; A = surface area, m^2; D_g = gas diffusivity, m^2/s; c = gas concentration, mol/m^3; and x = distance in the direction of diffusion, m. Applying Henry's law, the concentraion is given by $c = S.P$, where S = gas solubility, mol/(Pa.m^3); and P = gas partial pressure, Pa. Thus, $J = D_g.S.A.dP/dx$; and $D_g.S$ is known as the permeability coefficient, B, which is the (gas permeated) (film thickness) per unit time per unit packaging surface area and per unit pressure difference between environment and packaged material (i.e., mol/(s.Pa.m)).

Table 17.1 provides water-vapor permeability values of several packaging films. Table 17.2 gives gas permeability values of some packaging films.

Table 17.1 Permeability for Some Plastic Films at Room Temperature

Film	Permeability, $cm^3.mil.m^{-2}.day^{-1}.atm^{-1}$		
	Nitrogen	Oxygen	Carbon Dioxide
Saran	3	13	75
Nylon 6	25	100	400
Mylar (polyester)	20	80	260
Trithene or Kel-F	40	150	1,000
High-density polyethylene	700	2,000	10,000
Low-density polyethylene	3,500	12,000	70,000
Natural rubber	20,000	60,000	350,000
Silicone rubber		10^6	6×10^6

From M. Karel, O.R. Fennema and D.B. Lund, 1975, *Principles of Food Science, part II, Physical Principles of Food Preservation*, reprinted by courtesy of Marcel Dekker, Inc., New York.

Table 17.2 Permeability of Various Packaging Materials to Water Vapor at 100°F and 95% vs. 0% Relative Humidity

Material	Permeability Range, $g.mil\ 24\ h^{-1}\ 100\ in.^{-2}$
Plain cellophane	20-100
Nitrocellulose-coated cellophane	0.2-2.0
Saran-coated cellophane	0.1-0.5
Polyethylene, conventional	0.8-1.5
Polyethylene, low pressure	0.3-0.5
Saran	0.1-0.5
Vinyl-chloride-based films	0.5-8.0
Aluminum foil, 0.00035 in. thick	0.1-1.0
Aluminum foil, 0.0014 in. thick	< 0.1
Plastic paper foil laminations	< 0.1
Waxed papers	0.2-15.0
Coated papers	0.2-5.0
Mylar	0.8-1.5
(Poly)trifluorochloroethylene	0.01-0.1
Silicone rubber	> 200
Polypropylene	0.2-0.4

From M. Karel, O.R. Fennema and D.B. Lund 1975, *Principales of Food Science, part II, Physical Principles of Food Preservation*, reprinted by courtesy of Marcel Dekker, Inc., New York.

17.1.2. Permeability of Multilayer Material in Series or Parallel

Series. The arrangement of packaging films in series is shown in Fig. 17.1. The total permeability (B_T) is given by:

$$B_T = \frac{\Delta X_1 + \Delta X_2 + \Delta X_3}{\frac{\Delta X_1}{B_1} + \frac{\Delta X_2}{B_2} + \frac{\Delta X_3}{B_3}} \qquad (17.1)$$

This equation will not hold good for moisture- or gas-resistant coatings because the permeability of such material changes with coating thickness. This equation is valid when surface area of these films is constant, which is not true in spherical and cylindrical shaped packaging.

Parallel. The arrangement of packaging films in parallel is shown in Fig. 17.2. The total permeability (B_T) is given by:

$$B_T = \frac{A_1 . B_1 + A_2 . B_2 + A_3 . B_3}{A_1 + A_2 + A_3} \qquad (17.2)$$

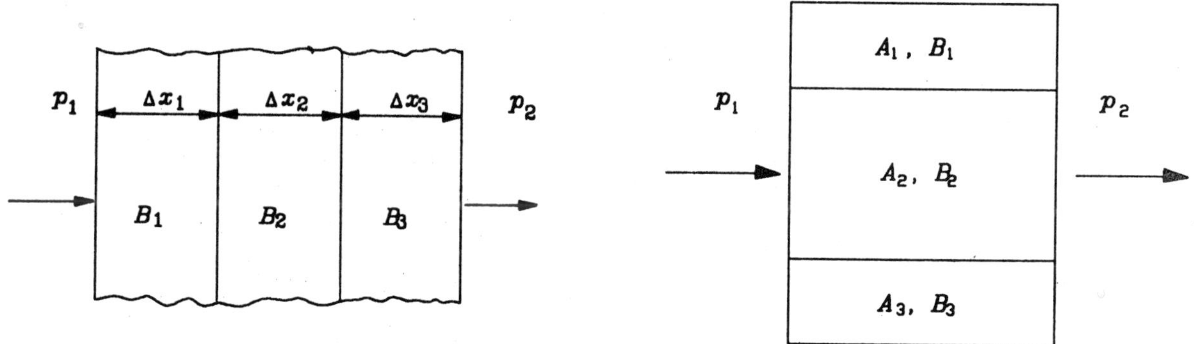

Fig. 17.1 Packaging films in series.

Fig. 17.2 Packaging films in parallel.

This is not true when thicknesses of the packaging materials also change.

17.1.3. Storage and Packaging

The changes in food environmental conditions during storage can be related to food and barrier properties of the material. To control moisture content of the food, the water vapor permeability of the packaging material can be selected based on the storage time (shelf life) required, initial moisture content of the food, and surface area and thickness of the packaging material. Similar approach is required to control the concentration of other gases such as O_2, CO_2, and N_2 within a packaged food environment. Respiration rates of fruits and vegetables can be reduced by lowering oxygen concentration in the packaging environment. Low O_2 concentration is also required to protect foods containing unsaturated fatty acids. For dried foods, controls of both O_2 and water vapor is essential for extended shelf life. For this purpose two films can be laminated to control O_2 and water vapor, one film controlling O_2 and, the other controlling water vapor transmission. Sometimes, to control water vapor and O_2 inside a packaged environment, desiccants, or oxygen scavengers can be placed inside a package.

17.1.4. Basics

Various packaging films are used in food packaging. For this lab three different films will be used, (1) waxed paper (7.5 mils thick), (2) Saran Wrap (1.0 mil thick), and (3) parchment paper (7.5 mils thick). A mil is equal to 0.001 in. Waxed papers vary widely in composition and structure, with a resultant wide range of water vapor transmission characteristics. Saran is a copolymer of poly vinyl chloride-poly vinylidene chloride (PVC-PVDC). Film characteristics are dependent on such factors as the degree of polymerization and molecular weight, spatial polymer orientation, presence of plasticizers (softeners), and the technique of film formation. Parchment paper is made with an acid-treated paper pulp. The acid treatment modifies the cellulose in the pulp and imparts water and oil resistance to the parchment. The water vapor transmission characteristics of these films are described by water vapor transmission rate, water-vapor permeance, and water vapor permeability. The water vapor transmission rate (WVTR) is

$$\text{WVTR} = 24.m_v \,/\, (t.A) \tag{17.3}$$

where m_v is mass gain or loss in g, t is time in h, and A is film surface area in m^2. WVTR is defined as the g of water vapor transmitted from 1 m^2 of film area in 24 h.

$$\text{Water vapor permeance} = \frac{\text{WVTR}}{\Delta p} = \frac{\text{WVTR}}{P_s(\text{RH}_1 - \text{RH}_2)} \tag{17.4}$$

where Δp is vapor pressure difference, mm Hg; P_s is saturation vapor pressure at 21°C, mm Hg; and RH_1 and RH_2 are relative humidities on each side of the film specimen. The water vapor permeance is defined as the gram of water vapor transmitted through 1 m^2 of film area in 24 h when the vapor pressure difference is 1 mm Hg.

$$\text{Water vapor permeability} = \frac{\text{water vapor permeance}}{\text{film thickness, cm}} \tag{17.5}$$

This is defined as a gram of water vapor permeated through 1 m^2 of film area in 24 h when vapor pressure difference is 1 mm Hg and film thickness is 1 cm. The Arrhenius equation provides the effect of temperature on the permeability coefficient, B:

$$B = B_o.\exp(-Ea/R.T) \tag{17.6}$$

where B_o = constant, Ea = activation energy, R = the gas constant, and T = the absolute temperature. The water vapor permeability of films increases exponentially with the increase in relative humidity due to water sorption and film swelling (Karel et al., 1975).

17.1.5. Permeability Measurement

Desiccant and water methods will be used for the measurements of water vapor permeability of packaging films, as explained in detail in Section 17.4. For the measurement of gas permeability the pressure increase method is easier to use. In the presure increase method, the test packaging film or membrane is mounted in a permeability cell, and constant pressure (P_1) of the test gas is applied on one side of the membrane, and on other side the increase in pressure (P_2) is noted with respect to time. The change in P_2 per unit time provides the permeability coefficient, B.

$$B = \frac{\Delta P_2}{\Delta t} \cdot \frac{V}{101,325} \cdot \frac{273}{T} \cdot \frac{\Delta X}{A} \tag{17.7}$$

Prelab Questions

Q1. What are some reasonable units for a packaging film mass transfer coefficient?

Q2. Why is the permeability of a package system different for wetted and unwetted inner surfaces?

17.2. OBJECTIVE

To determine the water vapor transmission characteristics of food packaging films.

17.3. APPARATUS

1. Relative humidity measuring system
2. Temperature measuring system
3. Glassware
4. Weighing balance
5. Controlled environment chamber

17.4. PROCEDURE

Procedures are the modifications of ASTM procedure E96-66 as described in the *ASTM Book of Standards* (1978). Two methods are used: the desiccant method (modification of ASTM procedure 9A) and the water method (modification of ASTM procedure 10B). The pressure increase method as shown in Fig. 17.3 can also be used.

Desiccant method: A desiccant (DRI-RITE) is placed in the bottom of a 140-mm diameter petri dish (Fig. 17.4). The package film is sealed over the opening of the dish, and the dish is placed in a controlled environment chamber (100% RH and 21°C temperature). Relative humidity inside the dish is assumed to be 0%. The sample dish is periodically weighed to determine moisture uptake. Record the data in Table 17.3.

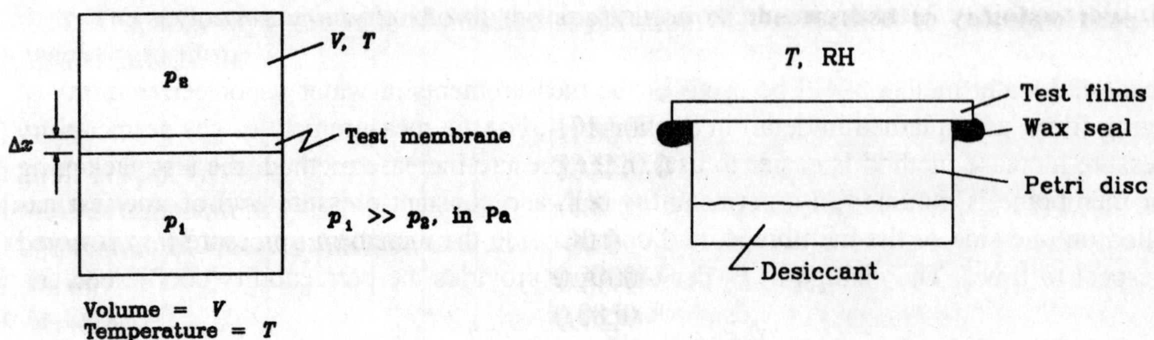

Fig. 17.3 Pressure increase method to measure film permeability for gases.

Fig. 17.4 Desiccant method to measure water permeability.

2. *Water method*: Water is placed in the bottom of a 140-mm diameter petri dish. The package film is sealed over the opening of the dish, and the dish is placed in a controlled environment chamber (30% RH and 21°C temperature). Relative humidity inside the dish is assumed to be 100%. The dish is periodically weighed to determine moisture loss through the film.

   Record the dish weight every day for about one week in Table 17.3. Handle with care as changes will be small.

17.5. RESULTS AND DISCUSSION

1. Plot weight gain or loss versus time for different films.

2. Calculate WVTR, water vapor permeance, and water vapor permeability for different films and compare with literature values.

Table 17.3 Data Sheet for Water Vapor Transmission Through Packaging Films

Petri dish diameter ________________ m, Film surface area ________________ m^2

Temperature (ambient) ______________ °C, Atmospheric pressure ______________ mm Hg

Relative humidity external (in water method) ________________________ %

Relative humidity external (in desiccant method) ______________________ %

Time, h	Desiccant Method Change in Weight (Δw), g			Water Method Change in Weight (Δw), g		
	Wax Paper	Saran Wrap	Parchment Paper	Wax Paper	Saran Wrap	Parchment Paper
1						
2						
3						
4						
5						
6						
7						

Film Thickness, mm

1. Wax paper ______________________
2. Saran wrap ______________________
3. Parchment paper ______________________

Prelab Questions for Laboratory 17

NAME:______________________________ DATE:________________________

ANSWER PRELAB QUESTIONS ON THIS SHEET

2.2.2. Thermophysical Properties

Finally, the program can calculate three thermophysical properties: (1) enthalpy, (2) specific heat, and (3) thermal conductivity. The results of these calculations are displayed in tabular form and are output on the same device chosen earlier.

The program is stored as [illegible]. To run the program, type PSYCO.

2.1.1. Example

Please indicate which two input variables you will use according to the following menu (two numbers separated by a comma) <RETURN>:

DRY BULB TEMPERATURE 1
WET BULB TEMPERATURE 2
RELATIVE HUMIDITY 3
DEW POINT TEMPERATURE 4
HUMIDITY RATIO 5
ENTHALPY 6
SPECIFIC VOLUME 7
? 2,3

Indicate all temperatures in degrees Celsius, relative humidity as a percentage, humidity ratio as kg/kg of dry air, enthalpy as kJ/kg of dry air, and specific volume as m^3/kg.

WHAT IS THE WET BULB TEMPERATURE? (AND RETURN) ? 25
WHAT IS THE RELATIVE HUMIDITY? (AND RETURN) ? 64

DRY BULB TEMP	WET BULB TEMP	DEW POINT TEMP	REL. HUM.	HUM. RATIO	SPEC. VOL.	ENTHALPY
(C)	(C)	(C)	(%)	(kg/kg)	(m^3/kg)	(kJ/kg)
30.07	25.00	22.75	64.0	[illegible]	[illegible]	84.8[illegible]

Would you like to see the output on the psychrometric chart? (Y or N) (and RETURN) Y

At the end of the execution [illegible] appear on the psychrometric chart.

Laboratory 18

PSYCHROMETRICS--USES AND APPLICATIONS

SUMMARY

This laboratory exercise provides background on various air-water-vapor mixture properties, various processes, and psychrometric chart. The application of psychrometric chart is discussed. Air properties will also be calculated using a computer program on a microcomputer.

18.1. BACKGROUND

18.1.1. Air Properties

A psychrometric chart provides various properties (i.e., enthalpy (h), relative humidity (RH), specific humidity (ω), specific volume (v), dew point temperature (T_{dp}) of air-water-vapor mixture at various dry (T_{db}) and wet bulb (T_{wb}) temperatures) at atmospheric pressure of 101,325 kPa. Thus, all the properties can be determined if any two properties are known. Air is a mixture of gases (mostly oxygen and nitrogen) and water vapor (very low pressure steam). Since the relative amount of vapor is small compared to gases, air is considered a mixture of perfect gases.

The aforementioned properties are explained as follows:

1. *Relative humidity (RH)*: defined as the ratio of the partial pressure of the water vapor (p_v) in the air to the saturation vapor pressure (p_g) in the air at the mixture temperature. Refers to the degree to which the air is saturated with water vapor.

2. *Specific humidity or humidity ratio* (ω): the ratio of the mass of vapor (m_v) in the air to the mass of dry air (m_a) for a given volume of mixture. It can be calculated as follows:

$$\omega = 0.6219\ (p_v/p_a) = 0.6219\ (\text{RH})(p_g/p_a) \tag{18.1}$$

where p_a is pressure of dry air.

3. *Dew point temperature* (T_{dp}): The dew point of a mixture of gas and vapor is the saturation temperature corresponding to the partial pressure of the vapor in the mixture or the temperature to which the mixture must be cooled at constant pressure to begin condensing the vapor. The humidity ratio remains constant during the cooling process. Cooling below the dew point will reduce both the vapor pressure and humidity ratio.

4. *Enthalpy (h)*: The enthalpy for a mixture containing a unit mass of dry air is:

$$h = C_{pa} \cdot T_{db} + \omega h_v, \text{ kJ/kg of dry air} \tag{18.2}$$

where C_{pa} is the specific heat at constant pressure for air (1.005 kJ/(kg.K), and h_v is the enthalpy of the vapor present within the air (i.e., the enthalpy of saturated vapor at T_{db}). h_v is $h_v = 2501.4 + 1.88\ (T_{db})$, taking 0°C as a reference temperature.

5. *Wet bulb temperature* (T_{wb}): This is nearly equal to the adiabatic saturation temperature of air. The air at T_{db} is saturated with water vapor without any heat transfer by evaporation of water from a large surface area. In practice, the adiabatic saturation device is approximated by wrapping a wick of porous gauze around the bulb of a thermometer or thermocouple, soaking the gauze in water, and forcing air to flow over and through the gauze, either with a fan or by swinging the thermometer through the air.

18.1.2. Various Processes on the Psychrometric Chart

1. *Sensible heating*: This occurs when the air flows over a surface that is warmer than itself, whereas ω remains unchanged. The heat transfer is equal to $h_2 - h_1$ (shown in Fig. 18.1).

2. *Sensible cooling*: The heat transfer is from the air to its surrounding. This occurs in a manner similar to sensible heating, except that the process moves in the opposite direction on the psychrometric chart. The final temperature should be above the T_{dp}.

3. *Cooling and dehumidification (Fig. 18.2)*: This occurs when the air is cooled below T_{dp}, thus causing moisture to condense from the air.

4. *Adiabatic cooling (Fig. 18.3)*: The cooling of air, without any heat transfer, by evaporation of water is adiabatic cooling. The air progresses from a low humidity at state 1 along the wet bulb temperature line to a cooler temperature and greater humidity at state 2. This generally occurs during the food drying process. As air gains moisture from the food, the humidity ratio increases.

5. *Heating and humidification*: This occurs in a cooling tower. The air picks heat and water vapor from the spraying water in the air stream (Fig. 18.4).

6. *Mixing of two air streams (Fig. 18.5)*: To calculate the properties of a mixture of two air streams, two air streams are first located on the psychrometric chart by knowing their two properties. Then, these two points are joined by astraight line, and point *C* (mixture) is located by dividing the straight line in reverse proportion to the masses of the individual air streams.

18.1.3. Mass and Energy Balances During Food Drying

Figure 18.6 shows the symbols for heat and mass balances for a food drying process where m_a = air flow rate (kg dry air/min), m_f = food flow rate (kg solids/min), M = food moisture content dry basis (kg water/kg solids), ω = humidity ratio of air (kg water vapor/kg dry air), T_f = food temperature (°C), h = enthalpy of air (kJ/kg dry air), and T_a = air temperature (°C).

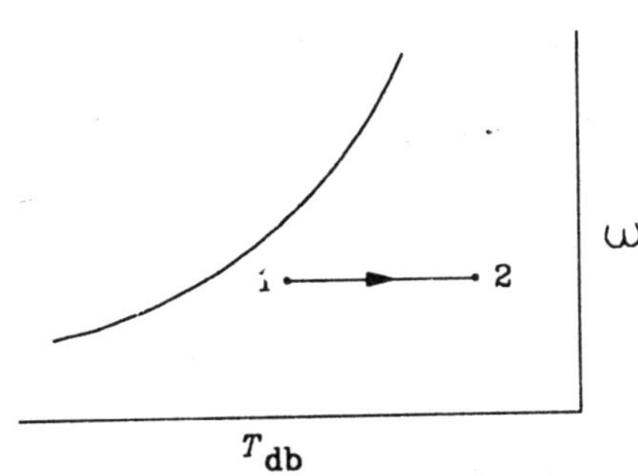

Fig. 18.1 Sensible heating process on a psychrometric chart.

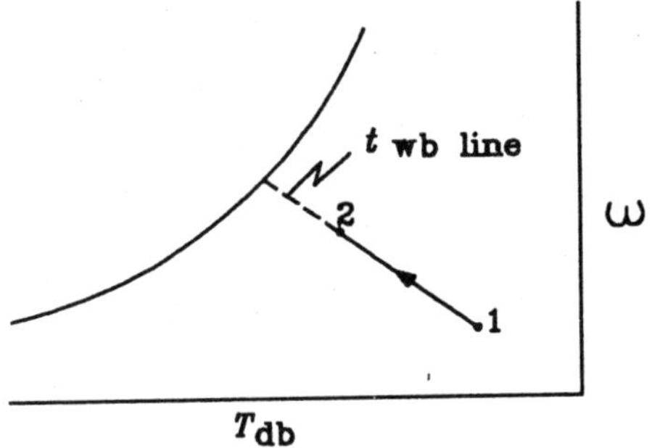

Fig. 18.3 **Adiabatic cooling on a psychrometric chart.**

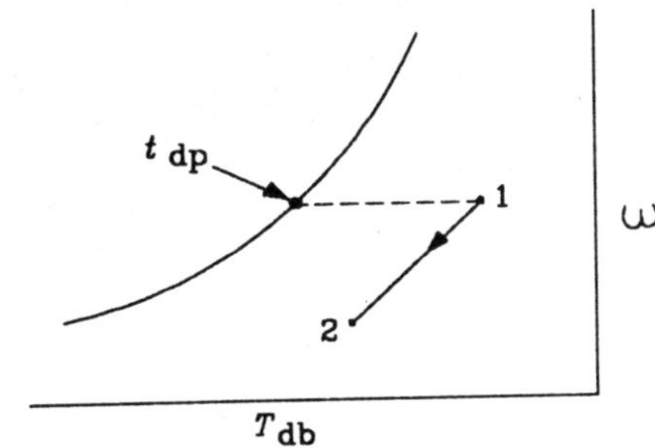

Fig. 18.2 Cooling and dehumidification on a psychrometric chart.

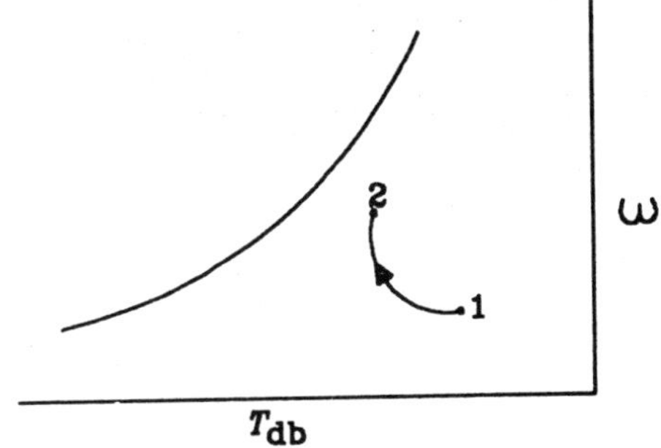

Fig. 18.4 **Heating and humidification process on a psychrometric chart.**

The moisture balance is given by

$$\dot{m}_a \cdot \omega_1 + \dot{m}_f \cdot M_1 = \dot{m}_a \cdot \omega_2 + \dot{m}_f \cdot M_2 \tag{18.3}$$

and the energy balance is given by

$$\dot{m}_a \cdot h_1 + \dot{m}_f (C_s + M_1 \cdot C_w)(T_{f1}) = \dot{m}_a \cdot h_2 + \dot{m}_f \cdot (C_s + M_2 \cdot C_w) \cdot T_{f2} + q \tag{18.4}$$

where C_s = specific heat of solids (kJ/(kg.K), C_w = specific heat of water, and q = heat losses from the dryer. Values h_1 and h_2 can be obtained from the psychrometric chart. Many drying system parameters can be established by using these equations and psychrometric chart.

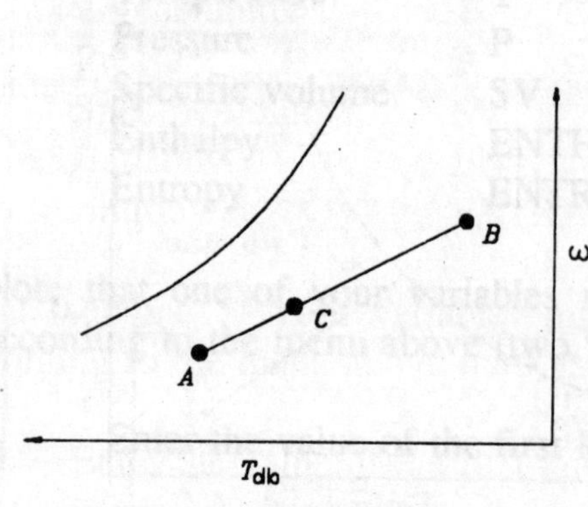

Fig. 18.5 Mixing of two air streams.

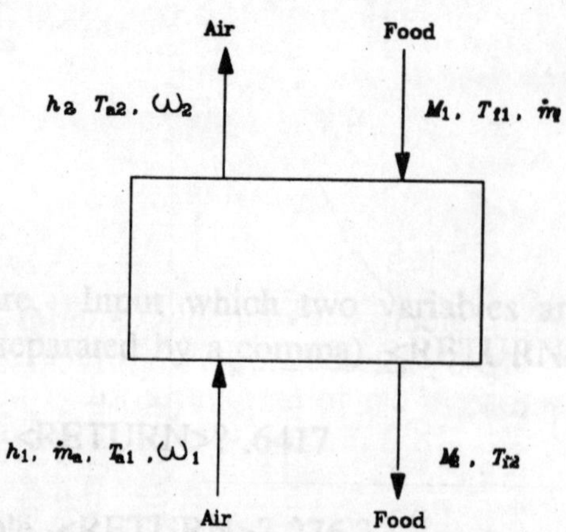

Fig. 18.6 Mass and energy balances for a food drying process.

Prelab Questions

Q1. Describe the relationship between humidity ratio and relative humidity.

Q2. Differentiate between dew point and wet bulb temperatures.

Q3. Differentiate between sensible and adiabatic coolings.

Q4. Differentiate between "cooling and humidification" and "heating and humidification."

18.2. APPLICATIONS OF PSYCHROMETRIC CHART

Solve the following problems using (1) a psychrometric chart, and (2) a microcomputer program PSYCO (see the Chapter on the Computer Programs).

1. Air has a dew point of 40°C and has a relative humidity of 60%. Determine the humidity ratio, the wet bulb temperature, and the dry bulb temperature.

2. Air in a smokehouse has a wet bulb temperature of 50°C and a dry bulb temperature of 70°C. Determine the relative humidity, the dew point, and the humidity ratio.

3. The wet bulb and dry bulb temperatures of the room air is 20°C and 25°C, respectively. Estimate the relative humidity, the enthalpy, and the specific volume of the air.

4. Ambient air at 20°C and 60% relative humidity is heated to 170°C. Determine the specific volume, the humidity ratio and the wet bulb temperature of the heated air.

5. The air is supplied to a process at a dry bulb temperature of 30°C and a relative humidity of 40% using cooling, dehumidification, and reheat sections. The air enters the cooling section at T_{db} = 35°C and RH = 60% at a flow rate of 1 m^3/s at 1 atm. Determine the rate of heat removed in the cooling and dehumidification section and the rate of heat addition in the reheat section.

6. The air is supplied to a device at a dry bulb temperature of 40°C and a relative humidity of 50% using a heating section at T_{db} = 10°C and RH = 80% at a flow rate of 1 m^3/s. In heating process, air dry bulb temperature is increased to 38°C. Then, in the humidifier section, the desired air conditions are achieved by spraying steam at 1 atm. Determine the steam temperature, rate of heat addition in heating section, and steam required in the humidifier section.

7. An air stream at T_{db1} = 10°C and RH_1 = 25% is mixed adiabatically at a flow rate of 0.8 m^3/s with a second air stream at a flow rate of 1 m^3/s at T_{db2} = 20°C and RH_2 = 50% at 1 atm. pressure. Calculate the flow rate, T_{db}, and RH of the mixture air stream.

8. In a food drying process, 3.2 kg/h of water is removed by passing air over the food material. Air at T_{db} = 20°C and RH = 30% is reheated to 60°C before passing to the dryer. The air leaves the dryer at T_{db} = 45°C and RH = 70%. Determine the required mass flow rate of dry air.

Prelab Questions for Laboratory 18

NAME:______________________________ **DATE:**____________________

ANSWER PRELAB QUESTIONS ON THIS SHEET

COMPUTER PROGRAMS

1. Notes in Using the Microcomputer

You should know the following about the microcomputer before you begin to use it. First of all, look on the far right side (or top) of the keyboard. There are 10 keys there with the numbers F1 through F10 on them (some may have 12 keys). There are also arrows and some words on these keys. Sometimes the computer will tell you that it will do something when you press PgUp or PgDn. It is referring to these keys when it does. To the left of the nine keys, there is a key with "enter" on it. This is sometimes referred to as the return key. You will sometimes be asked to type this, and you will also be required to type this after you have input a number as data to indicate to the computer that you are finished typing in the number.

Whenever you want to print out whatever is on the screen, you should press PrtSc. This will print out whatever is on the screen at the time. If there are any graphic plots on the screen, this could take a long time.

2. Programs Available

Enter "LOAD1"

2.1. Laboratory Programs

Rheology
Pipes and fittings
Pumps
Evaporators
Heat exchangers
Freezing
Thermal properties
Quit

Rheology: Does calculations for Newtonian fluids in a capillary rheometer and for power law fluids in a Brookfield-type rheometer.

Pipes and fittings: Does calculations for work and power requirements for a pump in a system consisting of pipes, tanks, various fittings with a constant friction coefficient, devices with a constant pressure drop, and changes in elevation.

Pumps: Does calculations for pump power requirements and net positive suction head considering only the friction from the suction and discharge pipes, and discharge pressure. Also calculates total head, horsepower, and efficiency for an experimental pump system where you measure suction and discharge pressures, flow rates, and power required.

Evaporators: Does calculations for steam required and plate area for a plate evaporator, which is single-effect, multieffect, or a mechanical vapor recompressed evaporator.

Heat exchangers: For several products of similar density, thermal conductivity, specific heat, and consistency coefficients, the program calculates the length of a heat exchanger and the overall heat transfer coefficient for each product, using the largest tubing size that will create turbulent flow in all of the products. The program then uses some products with properties similar to the products above, except for the fact that now n stays constant and m varies. The program then calculates the length of the heat exchangers and the overall heat transfer coefficient from the previous data except for the fact that the mass flow rate is such that the tubing size from the above calculation will create turbulent flow for all of the products.

Freezing: Calculates the freezing time required for a substance using either Nagaoka's or Plank's formula. It will also calculate the overall heat transfer coefficient from Nagaoka's or Plank's formula given the freezing time.

Thermal properties: Calculates heat capacity using formulas by Siebel and Charm and calculates heat capacity, density, thermal diffusivity, and thermal conductivity based on temperature and composition using formulas by Murakami.

2.1.1. Rheology Module

Rheology tutorial
Capillary rheometer
Brookfield rheometer
Calculate m and n from Brookfield rheometer
Quit

The rheology tutorial summarizes the material on rheology, which is given in the book in detail. A few multiple choice questions are given for more practice. If a wrong answer is entered, the correct answer will be provided by the computer.

A capillary rheometer asks the data (viscosity, density, and time) for two standards and then calculates the viscosity of the food based on its density and time required to flow through the capillary rheometer. The program can be terminated by entering "0" for fluid density.

A Brookfield rheometer calculates the apparent viscosity values based on the spindle number and dial reading.

The latter program calculates rheological parameters m and n for a fluid based on spindle speed and apparent viscosity data.

2.1.2. Pipes and Fittings

Pipes and fittings tutorial
Pipes and fittings calculations
Quit

The pipe and fittings tutorial provides description of the program and the related theory. A few multiple choice questions are also asked.

Pipe and fittings calculations calculates the fluid velocity and energy required based on the rheological coefficients m and n, pipe diameter, and flow rate (kg/s). After calculating the velocity, it asks the "type of calculation" with the following data:

1. Change of height
2. Change of pipe diameter
3. Device with pressure drop
4. Fittings with friction coefficient
5. Length of pipe
6. Quit

It then calculates the energy loss in pipes and fittings based on the data provided. If the choice is "4", then a list of 20 fittings is given or you can enter actual friction coefficient (if known). At the termination, it provides the "work done by the pump" and "power required at 100% efficiency."

2.1.3. Pump Performance Evaluation Module

Pump tutorial
Calculate work, power, and NPSH for simple system
Calculate pump curves for experimental pump systems
Quit

A pump tutorial explains the theory related to pump performance calculations. The details are given in the book. It also asks a few multiple choice questions.

The second part calculates work, power, and NPSH for simple pumping systems based on flow rate, pipe diameter, fluid density, discharge gauge pressure, discharge height, friction in suction line, friction in discharge line, pump efficiency, vapor pressure, and pump height.

The third part calculates pump curves for experimental pump system based on fluid density (lb/ft^3) and experimental data including input power (hp), suction head ($lb_f/in.^2$, gauge), discharge head ($lb_f/in.^2$, gauge), and flow rate (gal/min). For each data set, it provides total head and efficiency.

2.1.4. Evaporator Design Module

Evaporator tutorial
Single-effect evaporator
Single-effect evaporator with mechanical vapor recompression
Multieffect evaporator

The evaporator tutorial explains the theory related to single- and multieffect evaporators, and mechanical vapor recompression. It also asks a few multiple choice questions.

The single effect evaporator calculates steam required, outputs of product and vapor, product heat capacity, heat exchanger surface area for single-effect evaporator based on feed rate, solid concentrations of feed and product, specific heat of feed, enthalpy values of steam, vapor and condensate, temperatures of steam, product, and feed, and overall heat transfer coefficient supplied by the user.

The next section calculates product output, steam required, steam economy, and vapor flow-through compressor for a single- effect evaporator with mechanical vapor recompression based on feed rate; concentrations of feed and product; temperatures of feed and product; enthalpy values of steam, vapor; superheated enthalpy of final vapor at pressure of steam; compressor efficiency; and specific volume of inlet vapor.

The multiple effect evaporator calculates the amount of steam required, steam economy and surface area of heat exchanger based on number of stages; temperatures of steam, feed, and products; steam enthalpy; specific heats of feed and products; concentrations of feed and products; feed flow rate; and for each effect the values of overall heat transfer coefficient. It will also ask the values of the enthalpy of steam and condensate at various temperatures. Use program "Table" for the calculations of steam properties.

2.1.5. Heat Exchanger Design for Non-Newtonian Fluids

Heat exchanger tutorial
Heat exchanger calculations
Quit

The heat exchanger tutorial describes the background material related to various heat exchangers and their design. It also asks a few multiple choice questions at the end of the tutorial.

The heat exchanger calculations require the following inputs:

Number of data sets
Mass flow rate (kg/s)
Fluid density (kg/m^3)

Fluid heat capacity (J/(kg.K))
Thermal conductivity (W/(m.K))
Steam temperature (°C)
Product entry temperature (°C)
Flow behavior index, *n*, for each data set
Consistency coefficient (Pa.s^n)

and it provides the following outputs:

Length of heat exchanger
Diameter of heat exchanger
Convective heat transfer coefficient at various *n* values

2.1.6. Freezing Rate Calculations

Freezing tutorial
Calculates time from Plank's formula
Calculates time from Nagaoka'a formula
Calculates *h* given time from Plank's formula
Calculates *h* given time from Nagaoka'a formula
Quit

The freezing tutorial provides introductory material on the freezing rate calculations by Plank's and Nagaoka's equations. At the end of the session it asks a few multiple choice questions.

When the computer calculates time from Plank's formula it asks for the following inputs. Values are given to show its application.

Density	1050.0000
Latent heat of freezing	333220.0000
Freezing temperature	-1.8
Temperature of freezing medium	-30.0
P factor	0.3000
R factor	0.0850
Thickness	0.2500
Convective heat transfer coefficient	30.0000
Thermal conductivity	1.1080
Specific heat of unfrozen material	3520.0000
Specific heat of frozen material	2050.0000
Initial product temperature	5.0
Final product temperature	-10.0
Percent water (decimal)	0.7450
Packaging thickness	0.0400

Freezing time	0.0
Packaging thermal conductivity	0.6000

The output is as follows:

Time, Plank's formula	152541.6

Note: make sure that your units are consistent!

The following example shows the application of the method to calculate time from Nagaoka'a formula:

Density	1050.0
Latent heat of freezing	333220.0
Freezing temperature	-1.8
Temperature of freezing medium	-30.0
P factor	0.3000
R factor	0.0850
Thickness	0.2500
Convective heat transfer coefficient	30.0000
Thermal conductivity	1.1080
Specific heat of unfrozen material	3520.0000
Specific heat of frozen material	2050.0000
Initial product temperature	5.0
Final product temperature	-10.0
Percent water (decimal)	0.7450
Packaging thickness	0.0000
Freezing time	0.0
Packaging thermal conductivity	0.0000

The output is as follows:

Time, Nagaoka's formula	80869.2

Note: make sure that your units are consistent!

2.1.7. Thermal Properties Module Menu

Thermal properties tutorial
General formula for calculating heat capacity of food
Heat capacity from mass fractions of water, fat, and solids (Charm, 1978)
Heat capacity from percent water (Siebel, 1892)
Heat capacity from mass fractions and temperature (Murakami et al., 1985)
Density as previous
Thermal conductivity as previous, with volume fractions
Thermal diffusivity as previous
Quit

The thermal properties tutorial describes the background material. It explains Siebel, Charm, Heldman and Singh, and Muarkami et al. (1985) formulas for calculating thermal properties. At the end, it asks a few multiple choice questions.

The application of the Generalized formula for calculating heat capacity of food is shown as follows with an example:

Fraction of water	0.5
Fraction of fat	0.1
Fraction of protein	0.2
Fraction of carbohydrate	0.18
Fraction of ash	0.02
Heat capacity	2.844 kJ/(kg.K)

The heat capacity from fractions of water, fat, and solids (Charm, 1978) requires the following data:

Fraction of water	0.5
Fraction of fat	0.1
Fraction of solids	0.4

and the output is given as follows:

Heat capacity	2.759 kJ/(kg.K)

Similarly, the application of heat capacity from percent water (Siebel, 1892) is shown by the following example:

Fraction of water	0.5

and the ouput is:

Heat capacity	2.512 kJ/(kg.K)

The heat capacity from mass fractions of various things and temperature (Murakami) asks for the following data:

Mass fraction of protein	0.2
Mass fraction of fat	0.1
Mass fraction of carbohydrate	0.18
Mass fraction of fiber	0.02
Mass fraction of water	0.48
Mass fraction of ice	0.0
Mass fraction of air	0.0
Mass fraction of ash	0.02

and the output is:
Heat Capacity 3.151 kJ/(kg.K)

The following example shows the application of the calculation of the density using the Murakami et al. (1985) approach:

Temperature	20
Mass fraction of protein	0.2
Mass fraction of fat	0.1
Mass fraction of carbohydrate	0.18
Mass fraction of fiber	0.02
Mass fraction of water	0.48
Mass fraction of ice	0.2
Mass fraction of air	0.0
Mass fraction of ash	0.02

and the output is:
Density 1377.608 kg/m^3

The following example shows the application of the Murakami et al. (1985) approach to calculate thermal conductivity:

Temperature	25
Volume fraction of protein	0.2
Volume fraction of fat	0.1
Volume fraction of carbohydrate	0.18
Volume fraction of fiber	0.02
volume fraction of water	0.48
Volume fraction of ice	0.2
Volume fraction of air	0.02
Volume fraction of ash	0.02

and the output is as follows:
Thermal conductivity 0.864 W/(m.K)

The following example shows the application of the Murakami et al. (1985) approach to calculate the thermal diffusivity:

Temperature	30
Volume fraction of protein	0.2
Volume fraction of fat	0.1
Volume fraction of carbohydrate	0.18
Volume fraction of fiber	0.02
Volume fraction of water	0.48
Volume fraction of ice	0.15
Volume fraction of air	0.02

Volume fraction of ash 0.02
and the output is as follows:
Thermal Diffusivity 0.280 m^2/s

2.2. PSYCO: Psychrometric Calculations and Plotting

PSYCO's main function is to calculate the remaining five variables available from the psychrometric chart given any two of the seven possible variables. It will plot a psychrometric chart and display the calculated points on the chart. Finally, the program can calculate the thermophysical properties of the point in question.

User inputs to the main program are (1) two known input variables and their values, (2) output medium (screen, printer, or output file), and (3) to answer questions that are asked throughout the program run. Note that all temperatures should be entered in degrees Celsius, relative humidity as a percentage (not as a decimal between 0 and 1), humidity ratio as kg/kg of dry air, enthalpy as kJ/kg, and specific volume in m^3/kg.

The program will then access various subroutines depending on the input variables. All of the remaining variables contained on the psychrometric chart will be calculated when various equations are used.

There is a specific feature in this program: if the user presses the F1 key at any time during the program run, the program will stop and ask the user whether he/she would like to repeat the program. If he/she has made an error or would like to return to DOS, this feature is useful.

There are three options for an output display mode: (1) the output can be displayed on the screen, (2) the output can be displayed on the printer, and (3) the output can be stored in an output file. If the printer is chosen, the printer must be directly linked to the computer. The user will be asked at the beginning of the program which device he/she would like to use.

The results are displayed in tabular form at the end of the run.

2.2.1. Psychrometric Chart Subroutine

The data points can also be displayed on a psychrometric chart. This is accomplished by using a plotter. First, a labeled psychrometric chart is plotted using data points stored on the file DATA.DAT. The calculated points from the program are stored in PSY.DAT and recalled by the psychrometric chart subroutine. These points are plotted as triangles so that the user can recognize them on the chart. After six sets of data points have been plotted, the user can either continue plotting another six points on the same chart or plot a new chart containing the new points.

2.2.2. Thermophysical Properties

Finally, the program can calculate three thermal/physical properties: (1) viscosity, (2) specific heat, and (3) thermal conductivity. The results of these calculations are displayed in tabular form and are output on the same device chosen earlier.

The program is stored as PSYCO.EXE with XYLINE.DAT and RH.DAT. To run the program, type PYSCO.

Remember you must have the CAPS LOCK key on so that you are typing capital letters!!

2.2.3. Example

Please indicate which two input variables you will use according to the following menu (two numbers separated by a comma) <RETURN>.

DRY BULB TEMPERATURE	1
WET BULB TEMPERATURE	2
RELATIVE HUMIDITY	3
DEW POINT TEMPERATURE	4
HUMIDITY RATIO	5
ENTHALPY	6
SPECIFIC VOLUME	7

? 2,3

Indicate all temperatures in degrees Celsius, relative humidity as a percentage, humidity ratio as kg/kg of dry air, enthalpy as kJ/kg or J/g, and specific volume as m^3/kg.

WHAT IS THE WET BULB TEMPERATURE? <AND RETURN>? 25
WHAT IS THE RELATIVE HUMIDITY? <AND RETURN>? 64

DRY BULB TEMP (C)	WET BULB TEMP (C)	DEW BULB TEMP (C)	REL. HUM. (%)	HUM. RATIO (kg/kg)	SPEC. VOL. (m^3/kg)	ENTHALPY (kJ/kg)
30.02	25.00	22.45	64.0	0.0172	0.883	74.029

Would you like to see the output on the psychrometric chart? (Y or N) <and RETURN>?
Y

At the end of the complete run (i.e., after you do not repeat the program) the output will appear on the psychrometric chart.

Would you like to see the thermophysical properties for the run listed? (Y or N) <RETURN>? Y

THERMOPHYSICAL PROPERTIES

VISCOSITY (Pa.s)	SPECIFIC HEAT (kJ/(kg.K))	THERMAL CONDUCTIVITY (W/(m.K))
0.185D-04	1.053	0.02665

Would you like to repeat the program run (Y or N)?

2.3. TABLE

This is designed (1) to accept one known variable from the steam or freon table, (2) to accept two known variables, one always being pressure from the superheated/compressed liquid steam or freon-12 table, and (3) to output the remaining unknown variables based on these tables.

A menu showing the different choices available and the alphanumeric code associated with each variable is displayed. For example, the code for the temperature variable is T. The user must input a code indicating his/her desired selection.

The user must then input the numerical value of the variable selected. This number is assigned to the variable VALUE. Please note that temperatures must be input in degrees Celsius, pressure in $m^3/kg \times 10^3$. The program will then ask from which table (steam or freon) the user would like the information and the appropriate letter should be entered.

The program will search the appropriate table for the value input by the user. If the exact value is located, the complete set of eight numbers, temperature, pressure, specific volume (fluid and gas), enthalpy (fluid and gas), and entropy (fluid and gas) is stored in the array CYCLE(I). If the value is not found, the next value greater than the value lower and their entire rows are stored in the arrays CYCL1(I) and CYCL2(I). It will then interpolate between these arrays using a linear interpolation model. Since linearity is assumed, the interpolated values may not be as accurate as values calculated using other models.

2.3.1. Saturation Liquid

The output display consists of temperature, pressure, specific volume (fluid and gas), enthalpy (fluid and gas), and entropy (fluid and gas) values. The results will be displayed in tabular form at the end of the run. There is a repeat program question at the end of the run as well. At this point the user can either terminate the program and return to DOS or he/she can repeat the program for other variables and/or values.

2.3.2. Superheated/Compressed Liquid

A menu indicating the various choices available is displayed. The user must enter two variable codes (separated by a comma) indicating his/her selection in the order presented on the menu. One of the variables chosen must be pressure.

Numerical values of the two known values must now be input. The numerical values of the known variables must be entered in the same order as the variable code order. These numbers are assigned to variables VAL1 and VAL2, respectively.

Column numbers and known variable values are then assigned to the variables used in the search routine. These numbers are dependant on the codes initially chosen by the user.

In the search routine, the program first searches the table for the exact pressure value. If this is located in the data file, the routine will search for the exact second value (temperature, specific volume, enthalpy or entropy). If this number is located in the data file, the remainder of the row is stored in array CYCLE(DCOL).

If the exact second value is not found in the data file, the routine will locate the next value lower than the next value greater than the exact value and store their entire rows in CYCII2(0) AND CYCII3(0). It will then interpolate between the two arrays using a linear interpolation model.

If the exact pressure is not found, the routine will search the data table for the next value lower than the next value less than the initial known pressure. Within each of these two sets of pressures, the routine will then locate the next value less than the next value greater than the second known number. These four sets of values and their entire rows will be stored in arrays CYCII0(0), CYCII1(0), CYCII2(0), and CYCII3(0).

The program will then interpolate twice (1) using high and low values in each different pressure set separately, each interpolation value is stored as HII1 or HII2, SII1 or SII2, and VII1 or VII2 and (2) using the high and low values from the first interpolation and the pressure values.

The output displays temperature, pressure, specific volume, enthalpy, and entropy in tabular form at the completion of the run. There is a repeat question following the table display.

The F1 key can be pressed at any time during the program run and the program repeat question will appear. This feature is useful if an error has been made--the user can stop the run and try again.

The program contains TABLE.EXE, STEAM.DAT, FREON.DAT, SUPER.DAT and SUPFRO.DAT files. To run, type TABLE and press the return key.

REMEMBER, the CAPS LOCK key should be on so that you are typing CAPITAL LETTERS.

2.3.4. Example

This program will accept one known variable in the steam or freon table and output the remaining variables. It can also accept two known variables from the superheated/compressed liquid steam or freon table and output the remaining variables. Note that if you are using the superheated/compressed liquid tables, pressure must be one of your known variables. Press any key to continue. At any time during the program (run included) you can either repeat the program or end the program by pressing the F1 key <RETURN>. Input all temperatures in degrees Celsius, pressures in bars, enthalpy in kJ/kg, entropy in KJ/(kg.K), and specific volumes in $m^3/kg \times 10^3$. Press any key to continue. Indicate whether you are using the saturated tables (ST) or the superheated/compressed liquid tables (SCLT). <RETURN>? SCLT

Temperature	T
Pressure	P
Specific volume	SV
Enthalpy	ENTH
Entropy	ENTR

Note that one of your variables must be pressure. Input which two variables are known according to the foregoing menu. (two variable codes separated by a comma) <RETURN>? T,P

Enter the value of the first known variable. <RETURN>? 325

Enter the value of the second known variable. <RETURN>? 50

Indicate from which table, steam (S) or freon (F), you would like the information. <RETURN>? S

TEMPERATURE (C)	PRESSURE (bars)	SPECIFIC VOLUME ($m^3/kg \times 10^3$)	ENTHALPY (kJ/kg)	ENTROPY (kJ/(kg.K))
325.0	50.00	48.44250	2992.30	6.3178

Would you like to repeat the program: (Y or N)? Y

Indicate whether you are using the saturated tables (ST) or the superheated/compressed liquid tables (SCLT). <RETURN>? SCLT

Temperature	T
Pressure	P
Specific volume	SV
Enthalpy	ENTH
Entropy	ENTR

Note that one of your variables must be pressure. Input which two variables are known according to the menu above (two variable codes separated by a comma). <RETURN>? P,SV

Enter the value of the first known variable. <RETURN>? .6417

Enter the value of the second known variable. <RETURN>? 276.3

Indicate from which table, steam (S) or freon (F), you would like the information. <RETURN>? F

TEMPERATURE (C)	PRESSURE (bars)	SPECIFIC VOLUME ($m^3/kg \times 10^3$)	ENTHALPY (kJ/kg)	ENTROPY kJ/(kg.K))
10.0	0.64	276.29999	350.54	1.6539

Would you like to repeat the program? (Y or N)? Y

2.4. Thermal Processing

Type "RUN" to load the program. This will give you the following menu:

A. Create/edit data file
B. Celsius to Fahrenheit data conversion
C. Split data into heating/cooling portions
D. Data plotting program
E. Thermal process evaluation
X. Exit

Enter the letter of selection to execute the required task. It will provide you further information.

The first program will do the following functions.

Create new file, making it active
Make copy of file, making new copy active
Change active file
Junk first point in file
Take natural logarithm of x coordinates

Take natural logarithm of y coordinates
Take exponential of x coordinates
Take exponential of y coordinates
Take xnew = m*xold + b transformation on x coordinates
Take ynew = m*yold + b transformation on x coordinates
Divide file by n points
Quit

2.4.1. File Conversion from Celsius to Farenheit

Option "B" makes a copy of a data file containing time/temperature (°C) into a data file containing time/temperature (°F). It leaves the original data file unchanged.

2.4.2. Plotter

Option "D" allows you to plot a file or files. You are asked for two labels for each axis. Each label will appear on a separate line, in the order in which you enter them. Additional filenames will be plotted on the same graph as the first. If possible, try to use the filename FILE1.DAT when you create data files so as to save on disk space. If you need more than one file simultaneously, please use FILE2.DAT, then FILE3.DAT, and so forth. The plotter will only plot up to six files simultaneously.

Option 'E' provides the following menu:

A. Create/edit data file
B. Celsius to Fahrenheit data conversion
C. Split data into heating/cooling portions
D. Data plotting program
E. Thermal process evaluation
X. Exit

Enter the letter of selection for appropriate action.

Option "A" will provide the following menu:

Create new file, making it active
Make copy of file, making new copy active
Change active file
Junk first point in file
Take natural logarithm of x coordinates
Take natural logarithm of y coordinates
Take exponential of x coordinates
Take exponential of y coordinates
Take xnew = m*xold + b transformation on x coordinates

Take ynew = *m**yold + *b* transformation on y coordinates
Divide files by *n* points
Quit

Data Plotting Program--Initial Menu

Get data files from disk
*Disk Directory
Exit Program

Option 'E' provides the following menu:

Instructions
General method calculations
Formula method calculations
Exit to DOS

The first option provides the instruction on the use of these programs. To use the programs, you need to have your data in three separate data files: one file containing all of the raw data, one containing only the heating portion of the curve,and one containing only the cooling portion of the curve.

Your inital time/temperature data should be placed in an ASCII text file with the following format:

<label>: retort temperature, °F
<label>: cooling water temp, °F
<label>: initial temperature, °F
<label>: come-up time, min
<label>:

time0 (min)	temp0 (°F)
time1	temp1
etc.	etc.

The <label> field can be any words you want, *but each one MUST END WITH A COLON* (:). After the colon, enter a number. For example, on the first line, you might type the following:

Retort Temperature, °F: 250.0

If your temperature data happens to be in °C, after entering, rather than °F, run a program to convert to °F.

After creating your initial data file, you may create the heating and cooling files with your editor by making copies of your original and deleting the unnecessary points, or you can use the utility program SPLIT to do the job for you

SPLIT will allow you to graphically select where you would like to divide your original file into heating and cooling protions. It will then create two new data files (with names of your choice) containing only the points you specify as the heating and cooling portions of the curve, respectively.

You can analyze your data in one of several manners. The first is via the general method, which graphically calculates lethality based on your entire raw data file. It graphs your data, displays the data after transformation, visually indicates the integration area, and displays an F_o value calculated from your data.

Through use of the formula method, you can calculate F_o given B_b or vice versa for both simple or broken heating curves. To do this, you have to supply information about the heating process, such as j_h, f_h, f_2, x_{bh}, etc. depending on whether the heating curve is broken or simple. To obtain this information, you can regress yout heating and cooling data via the regression programs on the formula method menu. The information obtained from the regression will be written to a data file. It may then be used in the calculation of either B_b or F_o. The regression program will also aid you in deciding whether to treat your data as a simple or broken curve, both visually and through display of an R^2 (coefficient of determination) measure of data fitting.

Tabulating Your Results

For the programs that perform simple and broken heating curve calculations write their results out in tabular format to files named SIMPTAB.DAT and BROKTAB.DAT, respectively. This allows you to view the results of successive runs of the programs together as a whole. To see these files, enter the following at the DOS prompt:

C>type simptab.dat (or brokentab.dat) <return>

When you want to start the tables over (that is, erase all the computed enteries but leave the headings intact) type the following at the DOS prompt:

C>newtabls

2.4.3. Formula Method Menu

Regress raw heating data (time/temp)
Regress raw cooling data (time/temp)
Perform simple heating curve calculations
Perform broken heating curve calculations
Return to main menu

Thermal Process Evaluation by Ball's method

Calculate t_b given F_o
Calculate F_o given t_b
Exit program

These computer programs are available from the publisher on various diskette sizes. Please send the attached form to the publisher.

Request for Computer Programs

Disk size: 3.5 in. / 5.25 in. / 5.5 in.
double density / high density

Disk memory: 360 kbyte / 720 kbyte / 1.2 Mbyte / 1.44 Mbyte

Computer name and model:

Disk operating system:

Processor:

Name:

Mailing Address:

Please mail, with U.S. $ 5.00 for handling and postage, to:

Van Nostrand Reinhold
115 Fifth Avenue
New York, N.Y. 10003
U.S.A.

For Countries outside North America, contact the publisher for the procedure of getting the computer programs.

1. Acceleration

1 foot per second squared = 0.304 8 m/s^2

2. Area

1 square foot = 929.030 4 cm^2
1 square inch = 645.16 mm^2

3. Density

1 pound per cubic foot = 16.018 46 kg/m^3
1 pound per gallon = 99.776 37 kg/m^3
1 pound per gallon (U.S.) = 119.826 4 kg/m^3

4. Energy

1 British thermal unit (Btu) (International table) = 1.055 056 kJ
1 calorie (International table) = 4.186 8 J
1 foot pound-force = 1.355 818 J
1 kilowatt hour = 3.6 MJ

5. Force

1 dyne = 10 μN
1 kilogram-force = 9.806 65 N
1 poundal = 0.138 255 0 N
1 pound-force = 4.448 222 N

6. Heat

1 Btu foot per (square foot hour °F) = 1.730 735 W/(m.K)
1 Btu per cubic foot = 37.258 95 kJ/m^3
1 Btu per hour = 0.293 071 1 W
1 Btu per (pound °F) = 4.186 8 kJ/(kg.K)
1 Btu per (square foot hour °F) = 5.678 263 $W/(m^2.K)$

1 square foot hour °F per Btu	= 0.176 110 1 m².K/W

7. Length

1 foot	= 0.304 8 m
1 inch	= 25.4 mm

8. Mass

1 ounce (avoirdupois)	= 28.349 523 g
1 ounce (troy or apothecary)	= 31.103 476 8 g
1 pound (aviodupois)	= 0.453 592 37 kg
1 pound (troy or apothecary)	= 373.241 721 6 g
1 slug	= 14.593 90 kg

9. Power

1 Btu per hour	= 0.293 071 1 W
1 foot pound-force per hour	= 0.376 616 1 mW
1 horsepower (electric)	= 746 W

10. Pressure or stress (force per unit area)

1 atmosphere, standard (=760 torr)	= 101.325 kPa
1 bar	= 100 kPa
1 foot of water (39.2°F, 4°C)	= 2.988 98 kPa
1 mm mercury (conventional, 0 °C) or torr	= 133.322 4 Pa
1 poundal per square foot	= 1.488 164 Pa
1 pound-force per square inch (psi)	= 6.894 757 kPa

11. Temperature Intervals

1 degree Celcius	= 1 K
1 degree Fahrenheit	= 5/9 K
1 degree Rankine	= 5/9 K

12. Velocity (speed)

1 foot per second	= 304.8 mm/s
1 mile per hour	= 0.447 04 m/s

13. Viscosity

13.1 Viscosity, dynamic

1 poise	= 0.1 Pa.s
1 pound-force second per square foot	= 47.880 26 Pa.s

13.2 Viscosity, kinematic

1 square foot per second	= 92 903.04 mm^2/s
1 stokes	= 100 mm^2/s

14. Volume

1 bushel	= 36.368 72 dm^3
1 bushel (U.S. dry, 2 150.42 $in.^3$)	= 35.239 07 dm^3
1 cubic foot	= 28.316 85 dm^3
1 gallon	= 4.546 09 dm^3
1 gallon (U.S.)	= 3.785 412 dm^3

15. Volume rate of flow

1 cubic foot per second	= 28.316 85 dm^3/s
1 gallon per minute	= 75.768 17 cm^3/s
1 gallon (U.S.) per minute	= 63.090 20 cm^3/s

With the permission of the Canadian Standards Association, this material is reproduced from CAN/CSA-Z234.1-89 (Canadian Metric Practice Guide), which is copyrighted by CSA, 178 Rexdale Boulevard, Rexdale, Ontario, Canada, M9W 1R3. This copy of CSA Standard CAN/CSA-Z234.1-89 will not be updated to reflect amendments made to the original content of the CSA Standard after January, 1989. For up-to-date information, see the current edition of the CSA Catalogue of Standards.

Appendix B Steam Tables

Saturated Steam--Temperature Table

		Specific Volume, m³/kg		Internal Energy, kJ/kg			Enthalpy, kJ/kg			Entropy, kJ/kg		
Temp., °C T	Press., kPa P	Sat. Liquid v_f	Sat. Vapor v_g	Sat. Liquid u_f	Eavp. u_{fg}	Sat. Vapor u_g	Sat. Liquid h_f	Evap. h_{fg}	Sat. Vapor h_g	Sat. Liquid s_f	Evap. s_{fg}	Sat. Vapor s_g
0.01	0.61	0.001000	206.14	00.00	2375.3	2375.3	00.01	2501.3	2501.4	0.0000	9.1562	9.1562
5	0.87	0.001000	147.12	20.97	2361.3	2382.3	20.98	2489.6	2510.6	0.0761	8.9496	9.0257
10	1.23	0.001000	106.38	42.00	2347.2	2389.2	42.01	2477.7	2519.8	0.1510	8.7498	8.9008
15	1.70	0.001001	77.93	62.99	2333.1	2396.1	62.99	2465.9	2528.9	0.2245	8.5569	8.7814
20	2.34	0.001002	57.79	83.95	2319.0	2402.9	83.96	2454.1	2538.1	0.2966	8.3706	8.6672
25	3.17	0.001003	43.36	104.88	2304.9	2409.8	104.89	2442.3	2547.2	0.3674	8.1905	8.5580
30	4.25	0.001004	32.89	125.78	2290.8	2416.6	125.79	2430.5	2556.3	0.4369	8.0164	8.4533
35	5.63	0.001006	25.22	146.67	2276.7	2423.4	146.68	2418.6	2565.3	0.5053	7.8478	8.3531
40	7.38	0.001008	19.52	167.56	2262.6	2430.1	167.57	2406.7	2574.3	0.5725	7.6845	8.2570
45	9.59	0.001010	15.26	188.44	2248.4	2436.8	188.45	2394.8	2583.2	0.6387	7.5261	8.1648
50	12.35	0.001012	12.03	209.32	2234.2	2443.5	209.33	2382.7	2592.1	0.7038	7.3725	8.0763
55	15.76	0.001015	9.568	230.21	2219.9	2450.1	230.23	2370.7	2600.9	0.7679	7.2234	7.9913
60	19.94	0.001017	7.671	251.11	2205.5	2456.6	251.13	2358.5	2609.6	0.8312	7.0784	7.9096
65	25.03	0.001020	6.197	272.02	2191.1	2463.1	272.06	2346.2	2618.3	0.8935	6.9375	7.8310
70	31.19	0.001023	5.042	292.95	2176.6	2469.6	292.98	2333.8	2626.8	0.9549	6.8004	7.7553
75	38.58	0.001026	4.131	313.90	2162.0	2475.9	313.93	2321.4	2635.3	1.0155	6.6669	7.6824
80	47.39	0.001029	3.407	334.86	2147.4	2482.2	334.91	2308.8	2643.7	1.0753	6.5369	7.6122
85	57.83	0.001033	2.828	355.84	2132.6	2488.4	355.90	2296.0	2651.9	1.1343	6.4102	7.5445
90	70.14	0.001036	2.361	376.85	2117.7	2494.5	376.92	2283.2	2660.1	1.1925	6.2866	7.4791
95	84.55	0.001040	1.982	397.88	2102.7	2500.6	397.96	2270.2	2668.1	1.2500	6.1659	7.4159

Saturated Steam--Temperature Table (Continued)

		Specific Volume, m³/kg		Internal Energy, kJ/kg			Enthalpy, kJ/kg			Entropy, kJ/kg		
Temp., °C T	Pres., MPa P	Sat. Liquid v_f	Sat. Vapor v_g	Sat. Liquid u_f	Eavp. u_{fg}	Sat. Vapor u_g	Sat. Liquid h_f	Evap. h_{fg}	Sat. Vapor h_g	Sat. Liquid s_f	Evap. s_{fg}	Sat. Vapor s_g
100	0.10132	0.001044	1.6729	418.94	2087.6	2506.5	419.04	2257.0	2676.1	1.3069	6.0480	7.3549
105	0.12082	0.001048	1.4194	440.02	2072.3	2512.4	440.15	2243.7	2683.8	1.3630	5.9328	7.2958
110	0.14327	0.001052	1.2102	461.14	2057.0	2518.1	461.30	2230.2	2691.5	1.4185	5.8202	7.2387
115	0.16906	0.001056	1.0366	482.30	2041.4	2523.7	482.48	2216.5	2699.0	1.4734	5.7100	7.1833
120	0.19853	0.001060	0.8919	503.50	2025.8	2529.3	503.71	2202.6	2706.3	1.5276	5.6020	7.1296
125	0.2321	0.001065	0.7706	524.74	2009.9	2534.6	524.99	2188.5	2713.5	1.5813	5.4962	7.0775
130	0.2701	0.001070	0.6685	546.02	1993.9	2539.9	546.31	2174.2	2720.5	1.6344	5.3925	7.0269
135	0.3130	0.001075	0.5822	567.35	1977.7	2545.0	567.69	2159.6	2727.3	1.6870	5.2907	6.9777
140	0.3613	0.001080	0.5089	588.74	1961.3	2550.0	589.13	2144.7	2733.9	1.7391	5.1908	6.9299
145	0.4154	0.001085	0.4463	610.18	1944.7	2554.9	610.63	2129.6	2740.3	1.7907	5.0926	6.8833
150	0.4758	0.001091	0.3928	631.68	1927.9	2559.5	632.20	2114.3	2746.5	1.8418	4.9960	6.8379

Saturated Steam--Temperature Table (continued)

T	P	v_f	v_g	u_f	u_{fg}	u_g	h_f	h_{fg}	h_g	s_f	s_{fg}	s_g
155	0.5431	0.001096	0.3468	653.24	1910.8	2564.1	653.84	2098.6	2752.4	1.8925	4.9010	6.7935
160	0.6178	0.001102	0.3071	674.87	1893.5	2568.4	675.55	2082.6	2758.1	1.9427	4.8075	6.7502
165	0.7005	0.001108	0.2727	696.56	1876.0	2572.5	697.34	2066.2	2763.5	1.9925	4.7153	6.7078
170	0.7917	0.001114	0.2428	718.33	1858.1	2576.5	719.21	2049.5	2768.7	2.0419	4.6244	6.6663
175	0.8920	0.001121	0.2168	740.17	1840.0	2580.2	741.17	2032.4	2773.6	2.0909	4.5347	6.6256
180	1.0021	0.001127	0.19405	762.09	1821.6	2583.7	763.22	2015.0	2778.2	2.1396	4.4461	6.5857
185	1.1227	0.001134	0.17409	784.10	1802.9	2587.0	785.37	1997.1	2782.4	2.1879	4.3586	6.5465
190	1.2544	0.001141	0.15654	806.19	1783.8	2590.0	807.62	1978.8	2786.4	2.2359	4.2720	6.5079
195	1.3978	0.001149	0.14105	828.37	1764.4	2592.8	829.98	1960.0	2790.0	2.2835	4.1863	6.4698
200	1.5538	0.001157	0.12736	850.65	1744.7	2595.3	852.45	1940.7	2793.2	2.3309	4.1014	6.4323
205	1.7230	0.001164	0.11521	873.04	1724.5	2597.5	875.04	1921.0	2796.0	2.3780	4.0172	6.3952
210	1.9062	0.001173	0.10441	895.53	1703.9	2599.5	897.76	1900.7	2798.5	2.4248	3.9337	6.3585
215	2.104	0.001181	0.09479	918.14	1682.9	2601.1	920.62	1879.9	2800.5	2.4714	3.8507	6.3221
220	2.318	0.001190	0.08619	940.87	1661.5	2602.4	943.62	1858.5	2802.1	2.5178	3.7683	6.2861
225	2.548	0.001199	0.07849	963.73	1639.6	2603.3	966.78	1836.5	2803.3	2.5639	3.6863	6.2503
230	2.795	0.001209	0.07158	986.74	1617.2	2603.9	990.12	1813.8	2804.0	2.6099	3.6047	6.2146
235	3.06	0.001219	0.06537	1009.89	1594.2	2604.1	1013.62	1790.5	2804.2	2.6558	3.5233	6.1791
240	3.344	0.001229	0.05976	1033.21	1570.8	2604.0	1037.32	1766.5	2803.8	2.7015	3.4422	6.1437
245	3.648	0.001240	0.05471	1056.71	1546.7	2603.4	1061.23	1741.7	2803.0	2.7472	3.3612	6.1083
250	3.973	0.001251	0.05013	1080.39	1522.0	2602.4	1085.36	1716.2	2801.5	2.7927	3.2802	6.0730
255	4.319	0.001263	0.04598	1104.28	1496.7	2600.9	1109.73	1689.8	2799.5	2.8383	3.1992	6.0375
260	4.688	0.001276	0.04221	1128.39	1470.6	2599.0	1134.37	1662.5	2796.9	2.8838	3.1181	6.0019
265	5.081	0.001289	0.03877	1152.74	1443.9	2596.6	1159.28	1634.4	2793.6	2.9294	3.0368	5.9662
270	5.499	0.001302	0.03564	1177.36	1416.3	2593.7	1184.51	1605.2	2789.7	2.9751	2.9551	5.9301
275	5.942	0.001317	0.03279	1202.25	1387.9	2590.2	1210.07	1574.9	2785.0	3.0208	2.8730	5.8938
280	6.412	0.001332	0.03017	1227.46	1358.7	2586.1	1235.99	1543.6	2779.6	3.0668	2.7903	5.8571
285	6.909	0.001348	0.02777	1253.00	1328.4	2581.4	1262.31	1511.0	2773.3	3.1130	2.7070	5.8199
290	7.436	0.001366	0.02557	1278.92	1297.1	2576.0	1289.07	1477.1	2766.2	3.1594	2.6227	5.7821
295	7.993	0.001384	0.02354	1305.20	1264.7	2569.9	1316.30	1441.8	2758.1	3.2062	2.5375	5.7437
300	8.581	0.001404	0.02167	1332.00	1231.0	2563.0	1344.00	1404.9	2749.0	3.2534	2.4511	5.7045
305	9.202	0.001425	0.019948	1359.30	1195.9	2555.2	1372.40	1366.4	2738.7	3.3010	2.3633	5.6643
310	9.856	0.001447	0.018350	1387.10	1159.4	2546.4	1401.30	1326.0	2727.3	3.3493	2.2737	5.6230
315	10.547	0.001472	0.016867	1415.50	1121.1	2536.6	1431.00	1283.5	2714.5	3.3982	2.1821	5.5804
320	11.274	0.001499	0.015488	1444.60	1080.9	2525.5	1461.50	1238.6	2700.1	3.4480	2.0882	5.5362
330	12.845	0.001561	0.012996	1505.30	993.7	2498.9	1525.30	1140.6	2665.9	3.5507	1.8909	5.4417
340	14.586	0.001638	0.010797	1570.30	894.3	2464.6	1594.20	1027.9	2622.0	3.6594	1.6763	5.3357
350	16.513	0.001740	0.008813	1641.90	776.6	2418.4	1670.60	893.4	2563.9	3.7777	1.4335	5.2112
360	18.651	0.001893	0.006945	1725.20	626.3	2351.5	1760.50	720.5	2481.0	3.9147	1.1379	5.0526
370	21.03	0.002213	0.004925	1844.00	384.5	2228.5	1890.50	441.6	2332.1	4.1106	0.6865	4.7971
374.14	22.09	0.003155	0.003155	2029.60	0.0	2029.6	2099.30	0.0	2099.3	4.4298	0.0000	4.4298

From Van Wylen, G.J., and Sonntag, R.E., 1986. *Fundamentals of Classical Thermodynamics*. Wiley, New York. Reprinted with permission of John Wiley & Sons, Inc.

Superheated Vapor

	P = 0.010 MPa (45.81)				P = 0.050 MPa (81.33)				P = 0.10 MPa (99.63)			
T	*v*	*u*	*h*	*s*	*v*	*u*	*h*	*s*	*v*	*u*	*h*	*s*
Sat.	14.674	2437.9	2584.7	8.1502	3.240	2483.9	2645.9	7.5939	1.6940	2506.1	2675.5	7.3594
50	14.869	2443.9	2592.6	8.1749								
100	17.196	2515.5	2687.5	8.4479	3.418	2511.6	2682.5	7.6947	1.6958	2506.7	2676.2	7.3614
150	19.512	2587.9	2783.0	8.6882	3.889	2585.6	2780.1	7.9401	1.9364	2582.8	2776.4	7.6134
200	21.825	2661.3	2879.5	8.9038	4.356	2659.9	2877.7	8.1580	2.172	2658.1	2875.3	7.8343
250	24.136	2736.0	2977.3	9.1002	4.820	2735.0	2976.0	8.3556	2.406	2733.7	2974.3	8.0333
300	26.445	2812.1	3076.5	9.2813	5.284	2811.3	3075.5	8.5373	2.639	2810.4	3074.3	8.2158
400	31.063	2968.9	3279.6	9.6077	6.209	2968.5	3278.9	8.8642	3.103	2967.9	3278.2	8.5435
500	35.679	3132.3	3489.1	9.8978	7.134	3132.0	3488.7	9.1546	3.565	3131.6	3488.1	8.8342
600	40.295	3302.5	3705.4	10.1608	8.057	3302.2	3705.1	9.4178	4.028	3301.9	3704.7	9.0976
700	44.911	3479.6	3928.7	10.4028	8.981	3479.4	3928.5	9.6599	4.490	3479.2	3928.2	9.3398
800	49.526	3663.8	4159.0	10.6281	9.904	3663.6	4158.9	9.8852	4.952	3663.5	4158.6	9.5652
900	54.141	3855.0	4396.4	10.8396	10.828	3854.9	4396.3	10.0967	5.414	3854.8	4396.1	9.7767
1000	58.757	4053.0	4640.6	11.0393	11.751	4052.9	4640.5	10.2964	5.875	4052.8	4640.3	9.9764
1100	63.372	4257.5	4891.2	11.2287	12.674	4257.4	4891.1	10.4859	6.337	4257.3	4891.0	10.1659
1200	67.987	4467.9	5147.8	11.4091	13.597	4467.8	5147.7	10.6662	6.799	4467.7	5147.6	10.3463
1300	72.602	4683.7	5409.7	11.5811	14.521	4683.6	5409.6	10.8382	7.260	4683.5	5409.5	10.5183

	P = 0.20 MPa (120.23)				P = 0.30 MPa (133.55)				P = 0.40 MPa (143.63)			
T	*v*	*u*	*h*	*s*	*v*	*u*	*h*	*s*	*v*	*u*	*h*	*s*
Sat.	0.8857	2529.5	2706.7	7.1272	0.6058	2543.6	2725.3	6.9919	0.4625	2553.6	2738.6	6.8959
150	0.9596	2576.9	2768.8	7.2795	0.6339	2570.8	2761.0	7.0778	0.4708	2564.5	2752.8	6.9299
200	1.0803	2654.4	2870.5	7.5066	0.7163	2650.7	2865.6	7.3115	0.5342	2646.8	2860.5	7.1706
250	1.1988	2731.2	2971.0	7.7086	0.7964	2728.7	2967.6	7.5166	0.5951	2726.1	2964.2	7.3789
300	1.3162	2808.6	3071.8	7.8926	0.8753	2806.7	3069.3	7.7022	0.6548	2804.8	3066.8	7.5662
400	1.5493	2966.7	3276.6	8.2218	1.0315	2965.6	3275.0	8.0330	0.7726	2964.4	3273.4	7.8985
500	1.7814	3130.8	3487.1	8.5133	1.1867	3130.0	3486.0	8.3251	0.8893	3129.2	3484.9	8.1913
600	2.013	3301.4	3704.0	8.7770	1.3414	3300.8	3703.2	8.5892	1.0055	3300.2	3702.4	8.4558
700	2.244	3478.8	3927.6	9.0194	1.4957	3478.4	3927.1	8.8319	1.1215	3477.9	3926.5	8.6987
800	2.475	3663.1	4158.2	9.2449	1.6499	3662.9	4157.8	9.0576	1.2372	3662.4	4157.3	8.9244
900	2.706	3854.5	4395.8	9.4566	1.8041	3854.2	4395.4	9.2692	1.3529	3853.9	4395.1	9.1362
1000	2.937	4052.5	4640.0	9.6563	1.9581	4052.3	4639.7	9.4690	1.4685	4052.0	4639.4	9.3360
1100	3.168	4257.0	4890.7	9.8458	2.1121	4256.8	4890.4	9.6585	1.5840	4256.5	4890.2	9.5256
1200	3.399	4467.5	5147.3	10.0262	2.2661	4467.2	5147.1	9.8389	1.6996	4467.0	5146.8	9.7060
1300	3.630	4683.2	5409.3	10.1982	2.4201	4683.0	5409.0	10.0110	1.8151	4682.8	5408.8	9.8780

	P = 0.50 MPa (151.86)				P = 0.60 MPa (158.85)				P = 0.80 MPa (170.43)			
T	*v*	*u*	*h*	*s*	*v*	*u*	*h*	*s*	*v*	*u*	*h*	*s*
Sat.	0.3749	2561.2	2748.7	6.8213	0.3157	2567.4	2756.8	6.7600	0.2404	2576.8	2769.1	6.6628
200	0.4249	2642.9	2855.4	7.0592	0.3520	2638.9	2850.1	6.9665	0.2608	2630.6	2839.3	6.8158
250	0.4744	2723.5	2960.7	7.2709	0.3938	2720.9	2957.2	7.1816	0.2931	2715.5	2950.0	7.0384
300	0.5226	2802.9	3064.2	7.4599	0.4344	2801.0	3061.6	7.3724	0.3241	2797.2	3056.5	7.2328
350	0.5701	2882.6	3167.7	7.6329	0.4742	2881.2	3165.7	7.5464	0.3544	2878.2	3161.7	7.4089

Superheated Vapor (continued)

T	v	u	h	s	v	u	h	s	v	u	h	s
400	0.6173	2963.2	3271.9	7.7938	0.5137	2962.1	3270.3	7.7079	0.3843	2959.7	3267.1	7.5716
500	0.7109	3128.4	3483.9	8.0873	0.5920	3127.6	3482.8	8.0021	0.4433	3126.0	3480.6	7.8673
600	0.8041	3299.6	3701.7	7.3522	0.6697	3299.1	3700.9	8.2674	0.5018	3297.9	3699.4	8.1333
700	0.8969	3477.5	3925.9	8.5952	0.7472	3477.0	3925.3	8.5107	0.5601	3476.2	3942.2	8.3770
800	0.9896	3662.1	4156.9	8.8211	0.8245	3661.8	4156.5	8.7367	0.6181	3661.1	4155.6	8.6033
900	1.0822	3853.6	4394.7	9.0329	0.9017	3853.4	4394.4	8.9486	0.6761	3852.8	4393.7	8.8153
1000	1.1747	4051.8	4639.1	9.2328	0.9788	4051.5	4638.8	9.1485	0.7340	4051.0	4638.2	9.0153
1100	1.2672	4256.3	4889.9	9.4224	1.0559	4256.1	4889.6	9.3381	0.7919	4255.6	4889.1	9.2050
1200	1.3596	4466.8	5146.6	9.6029	1.1330	4466.5	5146.3	9.5185	0.8497	4466.1	5145.9	9.3855
1300	1.4521	4682.5	5408.6	9.7749	1.2101	4682.3	5408.3	9.6906	0.9076	4681.8	5407.9	9.5575

	P = 1.00 MPa (179.91)				P = 1.20 MPa (187.99)				P = 1.40 MPa (195.07)			
T	v	u	h	s	v	u	h	s	v	u	h	s
Sat.	0.19444	2583.6	2778.1	6.5865	0.16333	2588.8	2784.8	6.5233	0.14084	2592.8	2790.0	6.4693
200	0.2060	2621.9	2827.9	6.6940	0.16930	2612.8	2815.9	6.5898	0.14302	2603.1	2803.3	6.4975
250	0.2327	2709.9	2942.6	6.9247	0.19234	2704.2	2935.0	6.8294	0.16350	2698.3	2927.2	6.7467
300	0.2579	2793.2	3151.2	7.1229	0.2138	2789.2	3045.8	7.0317	0.18228	2785.2	3040.4	6.9534
350	0.2825	2875.2	3157.7	7.3011	0.2345	2872.2	3153.6	7.2121	0.2003	2869.2	3149.5	7.1360
400	0.3066	2957.3	3263.9	7.4651	0.2548	2954.9	3260.7	7.3774	0.2178	2952.5	3257.5	7.3026
500	0.3541	3124.4	3478.5	7.7622	0.2946	3122.8	3476.3	7.6759	0.2521	3121.1	3474.1	7.6027
600	0.4011	3296.8	3697.9	8.0290	0.3339	3295.6	3696.3	7.9435	0.2860	3294.4	3694.8	7.8710
700	0.4478	3475.3	3923.1	8.2731	0.3729	3474.4	3922.0	8.1881	0.3195	3473.6	3920.8	8.1160
800	0.4943	3660.4	4154.7	8.4996	0.4118	3659.7	4153.8	8.4148	0.3528	3659.0	4153.0	8.3431
900	0.5407	3852.2	4392.9	8.7118	0.4505	3851.6	4392.2	8.6272	0.3861	3851.1	4391.5	8.5556
1000	0.5871	4050.5	4637.6	8.9119	0.4892	4050.0	4637.0	8.8274	0.4192	4049.5	4636.4	8.7559
1100	0.6335	4255.1	4888.6	9.1017	0.5278	4254.6	4888.0	9.0172	0.4524	4254.1	4887.5	8.9457
1200	0.6798	4465.6	5145.4	9.2822	0.5665	4465.1	5144.9	9.1977	0.4855	4464.7	5144.4	9.1262
1300	0.7261	4681.3	5407.4	9.4543	0.6051	4680.9	5407.0	9.3698	0.5186	4680.4	5406.5	9.2984

	P = 1.60 MPa (201.41)				P = 1.80 MPa (207.15)				P = 2.00 MPa (212.42)			
T	v	u	h	s	v	u	h	s	v	u	h	s
Sat.	0.12380	2596.0	2794.0	6.4218	0.11042	2598.4	2797.1	6.3794	0.09963	2600.3	2799.5	6.3409
225	0.13287	2644.7	2857.3	6.5518	0.11673	2636.6	2846.7	6.4808	0.10377	2628.3	2835.8	6.4147
250	0.14184	2692.3	2919.2	6.6732	0.12497	2686.0	2911.0	6.6066	0.11144	2679.6	2902.5	6.5453
300	0.15862	2781.1	3034.8	6.8844	0.14021	2776.9	3029.2	6.8226	0.12547	2772.6	3023.5	6.7664
350	0.17456	2866.1	3145.4	7.0694	0.15457	2863.0	3141.2	7.0100	0.13857	2859.8	3137.0	6.9563
400	0.19005	2950.1	3254.2	7.2374	0.16847	2947.7	3250.9	7.1794	0.15120	2945.2	3247.6	7.1271
500	0.2203	3119.5	3472.0	7.5390	0.19550	3117.9	3469.8	7.4825	0.17568	3116.2	3467.6	7.4317
600	0.2500	3293.3	3693.2	7.8080	0.2220	3292.1	3691.7	7.7523	0.19960	3290.9	3690.1	7.7024
700	0.2794	3472.7	3919.7	8.0535	0.2482	3471.8	3918.5	7.9983	0.2232	3470.9	3917.4	7.9487
800	0.3086	3658.3	4152.1	8.2808	0.2742	3657.6	4151.2	8.2258	0.2467	3657.0	4150.3	8.1765
900	0.3377	3850.5	4390.8	8.4935	0.3001	3849.9	4390.1	8.4386	0.2700	3849.3	4389.4	8.3895
1000	0.3668	4049.0	4635.8	8.6938	0.3260	4048.5	4635.2	8.6391	0.2933	4048.0	4634.6	8.5901
1100	0.3958	4253.7	4887.0	8.8837	0.3518	4253.2	4886.4	8.8290	0.3166	4252.7	4885.9	8.7800
1200	0.4248	4464.2	5143.9	9.0643	0.3776	4463.7	5143.4	9.0096	0.3398	4463.3	5142.9	8.9607
1300	0.4538	4679.9	5406.0	9.2364	0.4034	4679.5	5405.6	9.1818	0.3631	4679.0	5405.1	9.1329

Superheated Vapor (continued)

T	P = 2.50 MPa (223.99) v	u	h	s	P = 3.00 MPa (233.90) v	u	h	s	P = 3.50 MPa (242.60) v	u	h	s
Sat.	0.07998	2603.1	2803.1	6.2575	0.06668	2604.1	2804.2	6.1869	0.05707	2603.7	2803.4	6.1253
225	0.08027	2605.6	2806.3	6.2639								
250	0.08700	2662.6	2880.1	6.4085	0.07058	2644.0	2855.8	6.2872	0.05872	2623.7	2829.2	6.1749
300	0.09890	2761.6	3008.8	6.6438	0.08114	2750.1	2993.5	6.5390	0.06842	2738.0	2977.5	6.4461
350	0.10976	2851.9	3126.3	6.8403	0.09053	2843.7	3115.3	6.7428	0.07678	2835.3	3104.0	6.6579
400	0.12010	2939.1	3239.3	7.0148	0.09936	2932.8	3230.9	6.9212	0.08453	2926.4	3222.3	6.8405
450	0.13014	3025.5	3350.8	7.1746	0.10787	3020.4	3344.0	7.0834	0,09196	3015.3	3337.2	7.0052
500	0.13998	3112.1	3462.1	7.3234	0.11619	3108.0	3456.5	7.2338	0.09918	3103.0	3450.9	7.1572
600	0.15930	3288.0	3686.3	7.5960	0.13243	3285.0	3682.3	7.5085	0.11324	3282.1	3678.4	7.4339
700	0.17832	3468.7	3914.5	7.8435	0.14838	3466.5	3911.7	7.7571	0.12699	3464.3	3908.8	7.6837
800	0.19716	3655.3	4148.2	8.0720	0.16414	3653.5	4145.9	7.9862	0.14056	3651.8	4143.7	7.9134
900	0.2159	3847.9	4387.6	8.2853	0.17980	3846.5	4385.9	8.1999	0.15402	3845.0	4384.1	8.1276
1000	0.2346	4046.7	4633.1	8.4861	0.19541	4045.4	4631.6	8.4009	0.16743	4044.1	4630.1	8.3288
1100	0.2532	4251.5	4884.6	8.6762	0.21098	4250.3	4883.3	8.5912	0.18080	4249.2	4881.9	8.5192
1200	0.2718	4462.1	5141.7	8.8569	0.22652	4460.9	5140.5	8.7720	0.19415	4459.8	5139.3	8.7000
1300	0.2905	4677.8	5404.0	9.0291	0.24206	4676.6	5402.8	8.9442	0.20749	4675.5	5401.7	8.8723

Compressed Liquid

T	P = 5.00 MPa (263.99) v	u	h	s	P = 10.00 MPa (311.06) v	u	h	s	P = 15.00 MPa (342.24) v	u	h	s
Sat.	0.0012859	1147.8	1154.2	2.9202	0.0014524	1393.0	1407.6	3.3596	0.0016581	1585.6	1610.5	3.6848
0	0.0009977	0.04	5.04	0.0001	0.0009952	0.09	10.04	0.0002	0.0009928	0.15	15.05	0.0004
20	0.0009995	83.65	88.65	0.2956	0.0009972	83.36	93.33	0.2945	0.0009950	83.06	97.99	0.2934
40	0.0010056	166.95	171.97	0.5705	0.0010034	166.35	176.38	0.5686	0.0010013	165.76	180.78	0.5666
60	0.0010149	250.23	255.30	0.8285	0.0010127	249.36	259.49	0.8258	0.0010105	248.51	263.67	0.8232
80	0.0010268	333.72	338.85	1.0720	0.0010245	332.59	342.83	1.0688	0.0010222	331.48	346.81	1.0656
100	0.0010410	417.52	422.72	1.3030	0.0010385	416.12	426.50	1.2992	0.0010361	414.74	430.28	1.2955
120	0.0010576	501.80	507.09	1.5233	0.0010549	500.08	510.64	1.5189	0.0010522	498.40	514.19	1.5145
140	0.0010768	586.76	592.15	1.7343	0.0010737	584.68	595.42	1.7292	0.0010707	582.66	598.72	1.7242
160	0.0010988	672.62	678.12	1.9375	0.0010953	670.13	681.08	1.9317	0.0010918	667.71	684.09	1.9260
180	0.0011240	759.63	765.25	2.1341	0.0011199	756.65	767.84	2.1275	0.0011159	753.76	770.50	2.1210
200	0.0011530	848.1	853.9	2.3255	0.0011480	844.5	856.0	2.3178	0.0011433	841.0	858.2	1.3104
220	0.0011866	938.4	944.4	2.5128	0.0011805	934.1	945.9	2.5039	0.0011748	929.9	947.5	2.4953
240	0.0012264	1031.4	1037.5	2.6979	0.0012187	1026.0	1038.1	2.6872	0.0012114	1020.8	1039.0	2.6771
260	0.0012749	1127.9	1134.3	2.8830	0.0012645	1121.1	1133.7	2.8699	0.0012550	1114.6	1133.4	2.8576
280					0.0013216	1220.9	1234.1	3.0548	0.0013084	1212.5	1232.1	3.0393
300					0.0013972	1328.4	1342.3	3.2469	0.0013770	1316.6	1337.3	3.2260
320									0.0014724	1431.1	1453.2	3.4247
340									0.0016311	1567.5	1591.9	3.6546

From Van Wylen, G.J., and Sonntag, R.E., 1986. *Fundamentals of Classical Thermodynamics*. Wiley, New York. Reprinted with permission of John Wiley & Sons, Inc.

Appendix C Refrigerant Tables

Saturated Refrigerant-12--Temperature Table

Temp., °C	Abs. Pres., MPa P	Specific Volume, m³/kg: Sat. Liquid v_f	Evap. v_{fg}	Sat. Vapor v_g	Internal Energy, kJ/kg: Sat. Liquid, u_f	Evap. u_{fg}	Sat. Vapor u_g	Enthalpy, kJ/kg: Sat. Liquid h_f	Evap. h_{fg}	Sat. Vapor h_g	Entropy, kJ/(kg.K): Sat. Liquid s_f	Evap. s_{fg}	Sat. Vapor s_g
-90	0.0028	0.000 608	4.414 937	4.415 545	-43.245	177.256	131.011	-43.243	189.618	146.375	-0.2084	1.0352	0.8268
-85	0.0042	0.000 612	3.036 704	3.037 316	-38.971	174.154	135.883	-38.968	187.608	148.640	-0.1854	0.9970	0.8116
-80	0.0062	0.000 617	2.137 728	2.138 345	-34.692	172.358	137.666	-34.688	185.612	150.924	-0.1630	0.9609	0.7979
-75	0.0088	0.000 622	1.537 030	1.537 651	-30.406	170.099	139.693	-30.401	183.625	153.224	-0.1411	0.9266	0.7855
-70	0.0123	0.000 627	1.126 654	1.127 280	-26.111	167.781	141.670	-26.103	181.640	155.536	-0.1197	0.8940	0.7744
-65	0.0168	0.000 632	0.840 534	0.841 166	-21.804	165.529	143.725	-21.793	179.651	157.857	-0.0987	0.8630	0.7643
-60	0.0226	0.000 637	0.637 274	0.637 910	-17.483	163.250	145.767	-17.469	177.653	160.134	-0.0782	0.8334	0.7552
-55	0.0300	0.000 642	0.490 358	0.491 000	-13.148	160.930	147.782	-13.129	175.641	162.512	-0.0581	0.8051	0.7470
-50	0.0391	0.000 648	0.382 457	0.383 105	- 8.797	158.658	149.861	-8.772	173.611	164.840	-0.0384	0.7779	0.7396
-45	0.0504	0.000 654	0.302 029	0.302 682	- 4.429	156.337	151.908	-4.396	171.558	167.163	-0.0190	0.7519	0.7329
-40	0.0642	0.000 659	0.241 251	0.241 910	- 0.042	153.990	153.948	-0.000	169.479	169.479	-0.0000	0.7269	0.7269
-35	0.0807	0.000 666	0.194 732	0.195 398	4.362	151.653	156.051	4.416	167.368	171.784	0.0187	0.7027	0.7214
-30	0.1004	0.000 672	0.158 703	0.159 375	8.787	149.288	158.875	8.854	165.222	174.076	0.0371	0.6795	0.7165
-25	0.1237	0.000 679	0.130 487	0.131 166	13.231	146.890	160.127	13.315	163.037	176.352	0.0552	0.6570	0.7121
-20	0.1509	0.000 685	0.108 162	0.108 847	17.697	144.488	162.185	17.800	160.810	178.610	0.0730	0.6352	0.7082
-15	0.1826	0.000 693	0.090 326	0.091 018	22.185	142.041	164.226	22.312	158.534	180.846	0.0906	0.6141	0.7046
-10	0.2191	0.000 700	0.075 946	0.076 640	22.698	139.567	166.265	26.851	156.207	183.058	0.1079	0.5936	0.7014
- 5	0.2610	0.000 708	0.064 255	0.064 963	31.235	137.053	168.288	31.420	153.823	185.243	0.1250	0.5736	0.6986
0	0.3086	0.000 716	0.054 673	0.055 389	35.801	134.503	170.304	36.022	151.376	187.397	0.1418	0.5542	0.6960
5	0.3626	0.000 724	0.046 761	0.047 485	40.396	131.904	172.300	40.659	148.859	189.518	0.1585	0.5351	0.6937
10	0.4233	0.000 733	0.040 180	0.040 914	45.027	129.256	174.283	45.337	146.265	191.602	0.1750	0.5165	0.6916
15	0.4914	0.000 743	0.034 671	0.035 413	49.693	126.549	176.242	50.058	143.586	193.644	0.1914	0.4983	0.6897
20	0.5673	0.000 752	0.030 028	0.030 780	54.401	123.779	178.180	54.828	140.812	195.641	0.2076	0.4803	0.6879
25	0.6516	0.000 763	0.026 091	0.026 854	59.156	120.932	180.088	59.653	137.933	197.586	0.2237	0.4626	0.6863
30	0.7449	0.000 774	0.022 734	0.023 508	63.962	118.002	181.964	64.539	134.936	199.475	0.2397	0.4451	0.6848
35	0.8477	0.000 786	0.019 853	0.020 641	68.828	114.974	183.802	69.494	131.805	201.299	0.2557	0.4277	0.6834
40	0.9607	0.000 798	0.017 373	0.018 171	73.755	111.839	185.594	74.527	128.525	203.051	0.2716	0.4101	0.6820

Saturated Refrigerant--Temperature Table (continued)

		Specific Volume, m^3/kg			Internal Energy, kJ/kg			Enthalpy, kJ/kg			Entropy, kJ/(kg.K)		
Temp., °C	Abs. Pres., MPa P	Sat. Liquid v_f	Evap. v_{fg}	Sat. Vapor v_g	Sat. Liquid, u_f	Evap. u_{fg}	Sat. Vapor u_g	Sat. Liquid h_f	Evap. h_{fg}	Sat. Vapor h_g	Sat. Liquid s_f	Evap. s_{fg}	Sat. Vapor s_g
45	1.0843	0.000 811	0.015 220	0.016 032	78.768	108.571	187.339	79.647	125.074	204.722	0.2875	0.3931	0.6806
50	1.2193	0.000 826	0.013 344	0.014 170	83.861	105.160	189.021	84.868	121.430	206.298	0.3034	0.3758	0.6792
55	1.3663	0.000 841	0.011 701	0.012 542	89.052	101.578	190.630	90.201	117.565	207.766	0.3194	0.3582	0.6777
60	1.5250	0.000 858	0.010 253	0.011 111	94.356	97.800	192.156	95.665	113.443	209.109	0.3355	0.3405	0.6760
65	1.6988	0.000 877	0.008 971	0.009 847	99.789	93.786	193.575	101.279	109.024	210.303	0.3518	0.3224	0.6742
70	1.8858	0.000 897	0.007 828	0.008 725	105.375	89.492	194.867	107.067	104.255	211.321	0.3683	0.3038	0.6721
75	2.0874	0.000 920	0.006 802	0.007 728	111.138	84.867	196.005	113.058	99.068	212.126	0.3851	0.2845	0.6697
80	2.3046	0.000 946	0.005 875	0.006 821	117.111	79.834	196.945	119.291	93.373	212.665	0.4023	0.2644	0.6667
85	2.5380	0.000 976	0.005 029	0.006 005	123.341	74.283	197.624	125.818	87.047	212.865	0.4201	0.2430	0.6631
90	2.7885	0.001 052	0.004 246	0.005 258	129.886	68.006	197.952	132.708	79.907	212.614	0.4385	0.2200	0.6585
95	3.0560	0.001 656	0.003 508	0.004 563	136.840	60.937	197.777	140.068	71.658	211.726	0.4579	0.1946	0.6526
100	3.3440	0.001 113	0.002 790	0.003 903	144.354	52.437	196.791	148.076	61.768	209.843	0.4788	0.1655	0.6444
105	3.6509	0.001 197	0.002 045	0.003 242	152.715	41.548	194.263	157.085	49.014	206.099	0.5023	0.1296	0.6319
110	3.9784	0.001 364	0.001 098	0.002 462	162.633	24.057	186.689	168.059	28.425	196.484	0.5322	0.0742	0.6064
115	4.1135	0.001 792	0.000 005	0.001 797	167.345	0.130	167.675	174.920	0.151	175.071	0.5651	0.0004	0.5655

Source: Gordon J. Van Wylen and Richard E. Sonntag, 1976, *Fundamentals of Classical Thermodynamics*, SI Version 3rd ed., Wiley, New York, Table A.3.1, pp. 670-671. Originally published by I.E. du Pont de Nemours & Company, Inc., 1955 and 1956. Reprinted by permission of John Wiley & Sons, Inc.

Saturated Refrigerant-12--Pressure Table

Pres., MPa, P	Temp. °C, T	Specific Volume, m³/kg: Sat. Liquid v_f	Evap. v_{fg}	Sat. Vapor v_g	Internal Energy, kJ/kg: Sat. Liquid, u_f	Evap. u_{fg}	Sat. Vapor u_g	Enthalpy, kJ/kg: Sat. Liquid h_f	Evap. h_{fg}	Sat. Vapor h_g	Entropy, kJ/(kg.K): Sat. Liquid s_f	Evap. s_{fg}	Sat. Vapor s_g
0.06	-41.42	0.0006578	0.2568	0.2575	-1.29	154.8	153.49	-1.25	170.19	168.90	-0.0054	0.7344	0.7290
0.10	-30.10	0.0006719	0.1593	0.1600	8.71	149.4	158.15	8.78	165.37	174.15	0.0368	0.6803	0.7171
0.12	-25.74	0.0006776	0.1342	0.1349	12.58	147.4	159.95	12.66	163.48	176.14	0.0526	0.6607	0.7133
0.14	-21.91	0.0006828	0.1161	0.1168	15.99	145.5	161.52	16.09	161.78	177.87	0.0663	0.6439	0.7102
0.16	-18.49	0.0006876	0.1024	0.1031	19.07	143.8	162.91	19.18	160.23	179.41	0.0784	0.6292	0.7076
0.18	-15.38	0.0006921	0.09156	0.09225	21.86	142.3	164.19	21.98	158.82	180.80	0.0893	0.6161	0.7054
0.20	-12.53	0.0006962	0.08284	0.08354	24.43	140.9	165.36	24.57	157.50	182.07	0.0992	0.6043	0.7035
0.24	- 7.42	0.0007040	0.06963	0.07033	29.06	138.4	167.44	29.23	155.09	184.32	0.1168	0.5836	0.7004
0.28	- 2.93	0.0007111	0.06005	0.06076	33.15	136.1	169.26	33.35	152.90	186.27	0.1321	0.5659	0.6980
0.30	1.11	0.0007177	0.05279	0.05351	36.85	134.0	170.88	37.08	150.92	188.00	0.1457	0.5503	0.6960
0.40	8.15	0.0007299	0.04248	0.04321	43.35	130.3	173.69	43.64	147.33	190.97	0.1691	0.5237	0.6928
0.50	15.60	0.0007438	0.03408	0.03482	50.30	126.3	176.61	50.67	143.35	194.02	0.1935	0.4964	0.6899
0.60	22.00	0.0007546	0.02837	0.02913	56.35	122.7	179.09	56.80	139.77	196.57	0.2142	0.4736	0.6878
0.70	27.65	0.0007686	0.02424	0.02501	61.75	119.5	181.23	62.29	136.45	198.74	0.2324	0.4536	0.6860
0.80	32.74	0.0007802	0.02110	0.02188	66.68	116.5	183.13	67.30	133.33	200.63	0.2487	0.4358	0.6845
0.90	37.37	0.0007914	0.01845	0.01942	71.22	113.6	184.81	71.93	130.36	202.29	0.2634	0.4198	0.6832
1.0	41.64	0.0008023	0.01664	0.01744	75.46	110.9	186.32	76.26	127.50	203.76	0.2770	0.4050	0.6820
1.2	49.31	0.0008237	0.01359	0.01441	83.23	105.7	188.95	84.21	122.03	206.24	0.3015	0.3784	0.6799
1.4	56.09	0.0008448	0.1138	0.01222	90.28	100.8	191.11	91.46	116.76	208.22	0.3232	0.3546	0.6778
1.6	62.19	0.0008660	0.00967	0.01054	96.80	96.2	192.95	98.19	111.62	209.81	0.3329	0.3429	0.6758

Source: K. Wark, 1983, *Thermodynamics*, 4th Ed., McGraw-Hill, New York, Table A-17M, p. 809,
Originally published by I.E. du Pont de Nemours & Company, Inc. 1969. Reproduced with permission of McGraw-Hill, Inc.

Superheated Refrigerant-12

Temp., °C	v, m³/kg	u, kJ/kg	h, kJ/kg	s, kJ/(kg.K)	v, m³/kg	u, kJ/kg	h, kJ/kg	s, kJ/(kg.K)	v, m³/kg	u, kJ/kg	h, kJ/kg	s, kJ/(kg.K)
	0.05 MPa (−45.16°C)				0.10 MPa (−30.10°C)				0.15 MPa (−20.14°C)			
−20.0	0.341 857	163.95	181.042	0.7912	0.167 701	163.09	179.861	0.7401				
−10.0	0.356 227	168.95	186.757	0.8133	0.175 222	168.18	185.707	0.7628				
0.0	0.370 508	174.04	192.567	0.8350	0.182 647	173.36	191.628	0.7849	0.114 716	167.41	184.619	0.7318
10.0	0.384 716	179.24	198.471	0.8562	0.189 994	178.63	197.628	0.8064	0.119 866	172.68	190.660	0.7543
20.0	0.398 863	184.53	204.469	0.8770	0.197 277	183.98	203.707	0.8275	0.124 932	178.02	196.762	0.7763
30.0	0.412 959	189.91	210.557	0.8974	0.204 506	189.42	209.866	0.8482	0.129 930	183.44	202.927	0.7977
40.0	0.427 012	195.38	216.733	0.9175	0.211 691	194.94	216.104	0.8684	0.134 873	188.93	209.160	0.8186
50.0	0.441 030	200.95	222.997	0.9372	0.218 839	200.54	222.421	0.8883	0.139 768	194.50	215.463	0.8390
60.0	0.455 017	206.59	229.344	0.9565	0.225 955	206.22	228.815	0.9078	0.144 625	200.14	221.835	0.8591
70.0	0.468 978	212.33	235.774	0.9755	0.233 044	211.98	235.285	0.9269	0.149 450	205.86	228.277	0.8787
80.0	0.482 917	218.14	242.282	0.9942	0.240 111	217.82	241.829	0.9457	0.154 247	211.65	234.789	0.8980
90.0	0.496 838	224.03	248.868	1.0126	0.247 159	223.73	248.446	0.9642	0.159 020	217.52	241.371	0.9169
									0.163 774	223.45	248.020	0.9354
	0.20 MPa (−12.53°C)				0.25 MPa (−6.25°C)				0.30 MPa (−0.84°C)			
0.0	0.088 608	171.95	189.669	0.7320	0.069 752	171.21	188.644	0.7139	0.057 150	170.44	187.585	0.6984
10.0	0.092 550	177.37	195.878	0.7543	0.073 024	176.71	194.969	0.7366	0.059 984	176.04	194.034	0.7216
20.0	0.096 418	182.85	202.135	0.7760	0.076 218	182.27	201.322	0.7587	0.062 734	181.67	200.490	0.7440
30.0	0.100 228	188.40	208.446	0.7972	0.079 350	187.88	207.715	0.7801	0.065 418	187.34	206.969	0.7658
40.0	0.103 989	194.02	214.814	0.8178	0.082 431	193.55	214.153	0.8010	0.068 049	193.07	213.480	0.7869
50.0	0.107 710	199.70	221.243	0.8381	0.085 470	199.27	220.642	0.8214	0.070 635	198.84	220.030	0.8075
60.0	0.111 397	205.46	227.735	0.8578	0.088 474	205.07	227.185	0.8413	0.073 185	204.67	226.627	0.8276
70.0	0.115 055	211.28	234.291	0.8772	0.091 449	210.92	233.785	0.8608	0.075 705	210.56	233.273	0.8473
80.0	0.118 690	217.17	240.910	0.8962	0.094 398	216.84	240.443	0.8800	0.078 200	216.51	239.971	0.8665
90.0	0.122 304	223.13	247.593	0.9149	0.097 327	222.83	247.160	0.8987	0.080 673	222.52	246.723	0.8853
100.0	0.125 901	229.16	254.339	0.9332	0.100 238	228.88	253.936	0.9171	0.083 127	228.59	253.530	0.0038
110.0	0.129 483	235.25	261.147	0.9512	0.103 134	234.99	260.770	0.9352	0.085 566	234.72	260.391	0.9220
	0.40 MPa (8.15°C)				0.50 MPa (15.60°C)				0.60 MPa (22.0°C)			
20.0	0.045 836	180.43	198.762	0.7199	0.035 646	179.11	196.935	0.6999				
30.0	0.047 971	186.24	205.428	0.7423	0.037 464	185.08	203.814	0.7230	0.030 422	183.86	202.116	0.7063
40.0	0.050 046	191.91	212.095	0.7639	0.039 214	191.05	210.656	0.7452	0.031 966	189.97	209.154	0.7291
50.0	0.052 072	197.95	218.779	0.7849	0.040 911	197.03	217.484	0.7667	0.033 450	196.07	216.141	0.7511
60.0	0.054 059	203.86	225.488	0.8054	0.042 565	203.03	224.315	0.7875	0.034 887	202.17	223.104	0.7723
70.0	0.056 014	209.82	232.230	0.8253	0.044 184	209.07	231.161	0.8077	0.036 285	208.29	230.062	0.7929
80.0	0.057 941	215.84	239.012	0.8448	0.045 774	215.14	238.031	0.8275	0.037 653	214.44	237.027	0.8129
90.0	0.059 846	221.90	245.837	0.8638	0.047 340	221.26	244.932	0.8467	0.038 995	220.61	244.009	0.8324
100.0	0.061 731	228.01	252.707	0.8825	0.048 886	227.43	251.869	0.8656	0.040 316	226.83	251.016	0.8514
110.0	0.063 600	234.18	259.624	0.9008	0.050 415	233.64	258.845	0.8840	0.041 619	233.08	258.053	0.8700
120.0	0.065 455	240.41	266.590	0.9187	0.051 929	239.90	265.862	0.9021	0.042 907	239.38	265.124	0.8882
130.0	0.067 298	246.69	273.605	0.9364	0.053 430	246.21	272.923	0.9198	0.044 181	245.72	272.231	0.9061

Superheated Refrigerant-12 (continued)

Temp., °C	v, m³/kg	u, kJ/kg	h, kJ/kg	s, kJ/(kg.K)	v, m³/kg	u, kJ/kg	h, kJ/kg	s, kJ/(kg.K)	v, m³/kg	u, kJ/kg	h, kJ/kg	s, kJ/(kg.K)
	0.70 MPa (27.65°C)				0.80 MPa (32.74°C)				0.90MPa (37.37°C)			
40.0	0.026 761	188.85	207.580	0.7148	0.022 830	187.66	205.924	0.7016	0.019 744	186.40	204.170	0.6982
50.0	0.028 100	195.08	214.745	0.7373	0.024 068	194.04	213.290	0.7248	0.020 912	192.94	211.765	0.7131
60.0	0.029 387	201.28	221.854	0.7590	0.025 247	200.36	220.558	0.7469	0.022 012	199.40	219.212	0.7358
70.0	0.030 632	207.49	228.931	0.7799	0.026 380	206.66	227.766	0.7682	0.023 062	205.81	226.564	0.7575
80.0	0.031 843	213.71	235.997	0.8002	0.027 477	212.96	234.941	0.7888	0.024 072	212.19	233.856	0.7785
90.0	0.033 027	219.95	243.066	0.8199	0.028 545	219.27	242.101	0.8088	0.025 051	218.57	241.113	0.7987
100.0	0.034 189	226.21	250.146	0.8392	0.029 588	225.59	249.260	0.8283	0.026 005	224.91	248.355	0.8184
110.0	0.035 332	232.51	257.247	0.8579	0.030 612	231.94	256.428	0.8472	0.026 937	231.35	255.593	0.8376
120.0	0.036 458	238.85	264.374	0.8763	0.031 619	238.32	263.613	0.8657	0.027 851	237.77	262.839	0.8562
130.0	0.037 572	245.23	271.531	0.8943	0.032 612	244.73	270.820	0.8838	0.028 751	244.22	270.100	0.8745
140.0	0.038 673	251.65	278.720	0.9119	0.033 592	251.18	278.055	0.9016	0.029 639	250.71	277.381	0.8923
150.0	0.039 764	258.11	285.946	0.9292	0.034 563	257.67	285.320	0.9189	0.030 515	257.22	284.687	0.9098
	1.00 MPa (41.64°C)				1.20 MPa (49.31°C)				1.40 MPa (56.09°C)			
50.0	0.018 366	191.80	210.162	0.7021	0.014 483	189.28	206.661	0.6812				
60.0	0.019 410	198.40	217.810	0.7254	0.015 463	196.25	214.805	0.7060	0.012 579	193.85	211.457	0.6876
70.0	0.020 397	204.92	225.319	0.7476	0.016 368	203.05	222.687	0.7293	0.013 448	200.99	219.822	0.7123
80.0	0.021 341	211.40	232.739	0.7689	0.017 221	209.73	230.398	0.7514	0.014 247	207.95	227.891	0.7355
90.0	0.022 251	217.85	240.101	0.7895	0.018 032	216.36	237.995	0.7727	0.014 997	214.77	235.766	0.7575
100.0	0.023 133	224.30	247.430	0.8094	0.018 812	222.94	245.518	0.7931	0.015 710	221.52	243.512	0.7785
110.0	0.023 993	230.75	254.743	0.8287	0.019 567	229.51	252.993	0.8129	0.016 393	228.22	251.170	0.7988
120.0	0.024 835	237.22	262.053	0.8475	0.020 301	236.08	260.441	0.8320	0.017 053	234.90	258.770	0.8183
130.0	0.025 661	243.71	269.369	0.8659	0.021 018	242.65	267.875	0.8507	0.017 695	241.56	266.334	0.8373
140.0	0.026 474	250.23	276.699	0.8839	0.021 721	249.24	275.307	0.8689	0.018 321	248.23	273.877	0.8558
150.0	0.027 275	256.77	284.047	0.9015	0.022 412	255.85	282.745	0.8867	0.018 934	254.90	281.411	0.8738
160.0	0.028 068	263.35	291.419	0.9187	0.023 093	262.48	290.195	0.9041	0.019 535	261.60	288.946	0.8914
	1.60 MPa (62.19°C)				1.80 MPa (67.77°C)				2.00 MPa (72.89°C)			
70.0	0.011 208	198.72	216.650	0.6959	0.009 406	196.12	213.049	0.6794				
80.0	0.011 984	206.00	225.177	0.7204	0.010 187	203.86	222.198	0.7057	0.008 704	201.45	218.859	0.6909
90.0	0.012 698	213.07	233.390	0.7433	0.010 844	211.24	230.835	0.7298	0.009 406	209.24	228.056	0.7166
100.0	0.013 366	220.01	241.397	0.7651	0.011 526	218.41	239.155	0.7524	0.010 035	216.69	236.760	0.7402
110.0	0.014 000	226.86	249.264	0.7859	0.012 126	225.44	247.264	0.7739	0.010 615	223.92	245.154	0.7624
120.0	0.014 608	233.66	257.035	0.8059	0.012 697	232.37	255.228	0.7944	0.011 159	231.02	253.341	0.7835
130.0	0.015 195	240.43	264.742	0.8253	0.013 244	239.25	263.094	0.8141	0.011 676	238.03	261.384	0.8037
140.0	0.015 765	247.18	272.406	0.8440	0.013 772	246.10	270.891	0.8332	0.012 172	244.98	269.327	0.8232
150.0	0.016 320	253.93	280.044	0.8623	0.014 284	252.93	278.642	0.8518	0.012 651	251.90	277.201	0.8420
160.0	0.016 864	260.69	287.669	0.8801	0.014 784	259.75	286.364	0.8698	0.013 116	258.80	285.027	0.8603
170.0	0.017 398	267.45	295.290	0.8975	0.015 272	266.58	294.069	0.8874	0.013 570	265.68	292.822	0.8781
180.0	0.017 923	274.24	302.914	0.9145	0.015 752	273.41	301.767	0.9046	0.014 013	272.57	300.598	0.8955

Superheated Refrigerant-12 (continued)

Temp., °C	v, m³/kg	u, kJ/kg	h, kJ/kg	s, kJ/(kg.K)	v, m³/kg	u, kJ/kg	h, kJ/kg	s, kJ/(kg.K)	v, m³/kg	u, kJ/kg	h, kJ/kg	s, kJ/(kg.K)
	2.50 MPa (84.20°C)				3.00 MPa (93.99°C)				3.50 MPa (102.60°C)			
90.0	0.006 595	203.07	219.562	0.6823								
100.0	0.007 264	211.69	229.852	0.7103	0.005 231	204.84	220.529	0.6770				
110.0	0.007 837	219.68	239.271	0.7352	0.005 886	214.41	232.068	0.7075	0.004 324	206.99	222.121	0.6750
120.0	0.008 351	227.31	248.192	0.7582	0.006 419	222.95	242.208	0.7336	0.004 959	217.52	234.875	0.7078
130.0	0.008 827	234.73	256.794	0.7798	0.006 887	230.97	251.632	0.7573	0.005 456	226.57	245.661	0.7349
140.0	0.009 273	242.00	265.180	0.8003	0.007 313	238.68	260.620	0.7793	0.005 884	234.93	255.524	0.7591
150.0	0.009 697	249.17	273.414	0.8200	0.007 709	246.19	269.319	0.8001	0.006 270	242.90	264.846	0.7814
160.0	0.010 104	256.28	281.540	0.8390	0.008 083	253.57	277.817	0.8200	0.006 626	250.63	273.817	0.8023
170.0	0.010 497	263.35	289.589	0.8574	0.008 439	260.85	286.171	0.8391	0.006 961	258.18	282.545	0.8222
180.0	0.010 879	270.39	297.583	0.8752	0.008 782	268.08	294.422	0.8575	0.007 279	265.62	291.100	0.8413
190.0	0.011 250	277.42	305.540	0.8926	0.009 114	275.26	302.597	0.8753	0.007 584	272.98	299.528	0.8597
200.0	0.011 614	284.44	313.472	0.9095	0.009 436	282.41	310.718	0.8927	0.007 878	280.29	307.864	0.8775
	4.00 MPa (110.25°C)											
120.0	0.003 736	209.92	224.863	0.6771								
130.0	0.004 325	221.14	238.443	0.7111								
140.0	0.004 781	230.58	249.703	0.7386								
150.0	0.005 172	239.22	259.904	0.7630								
160.0	0.005 522	247.40	269.492	0.7854								
170.0	0.005 845	255.30	278.684	0.8063								
180.0	0.006 147	263.01	287.602	0.8262								
190.0	0.006 434	270.59	296.326	0.8453								
200.0	0.006 708	273.07	304.906	0.8636								
210.0	0.006 972	285.49	313.380	0.8813								
220.0	0.007 228	292.86	321.774	0.8985								
230.0	0.007 477	300.20	330.108	0.9152								

Source: Gordon J. Van Wylen and Richard E. Sonnntag, 1976, *Fundamentals of Classical Thermodynamics*, SI Version 3d ed., Wiley, New York, pp. 672-675, Table A.3.2. Originally published by E.I. du Pont de Nemours & Company, Inc., 1955 and 1956. Reprinted by permission of John Wiley & Sons, Inc.

Appendix D -- Psychrometric Chart (reproduced with permission of Carrier Corporation, Syracuse, New York)

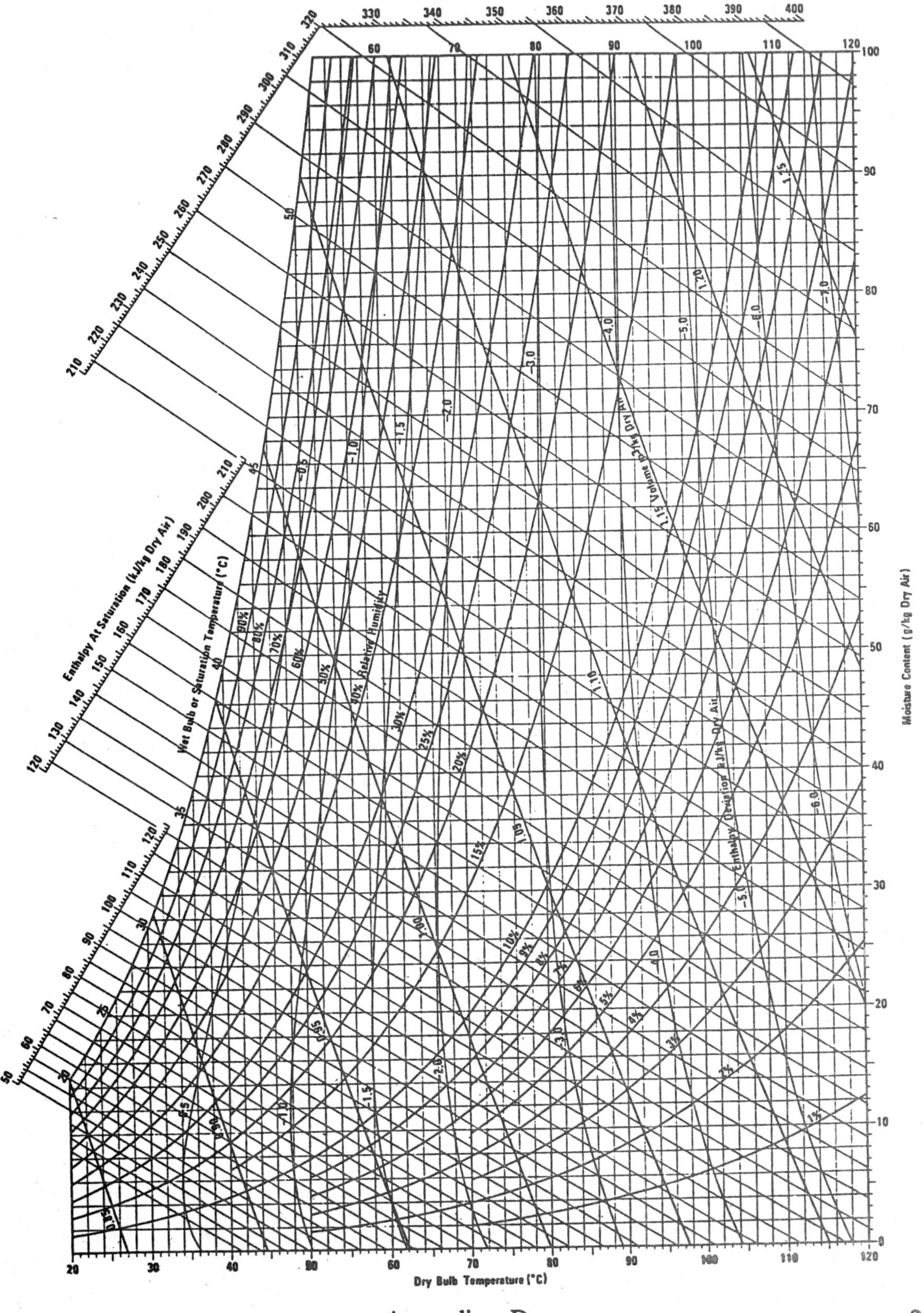

Appendix E--References

Anonymous. 1990. Food technology style guide. *Food Technology* 44(7):129

ASHRAE. 1981. *ASHRAE Handbook of Fundamentals*. American Society of Heating Refrigerating and Air-Conditioning Engineers, Inc., Atlanta, Ga.

ASTM. 1978. *Annual Book of ASTM Standards*. American Society of Testing and Materials, Philadelphia, PA.

Bate-Smith, F.C., Lea, C.H., and Sharp, J.G. 1943. Dried meat. *Soc. Chem. Ind.* (London) 62T:100-104.

Boucher, D.F., and Alves, G.E. 1973. Fluid and particle mechanics. In *Chemical Engineers' Handbook*, R.H. Perry and C.H. Chilton (eds.). McGraw-Hill Book Co., New York.

Canadian Standards Association. 1989. CSA Standard Z 234.1-89. *Canadian Metric Practice Guide*. Canadian Standards Assoc., Rexdale, Ontario, Canada.

Charm, S.E. 1978. *Fundamentals of Food Engineering*, AVI Publishing Co., Westport, CT.

Cleland, A.C., and Earle, R.L. 1979. Prediction of freezing times for foods in rectangular packages. *J. Food Sci.* 44:964.

Cleland, D.J., Cleland, A.C., and Earle, R.L. 1987. Prediction of freezing and thawing times for multidimensional shapes by simple formulae. Part I and II. *Int. J. Refrigeration*, 156-164 and 234-240.

Dickerson, R.W. 1965. An apparatus for the measurement of thermal diffusivity of foods. *Food Technol.* 19(5):198-204.

Dodge, D.W., and Metzner, A.B. 1959. Turbulent flow of non-Newtonian systems. *AIChE J.* 5:189, Errata *AIChE J.* 8:143 (1962).

Ede, A.J. 1949. The calculation of the freezing and thawing of foodstuffs. *Mod. Refrig.* 52:52.

Fields, M.L. 1977. *Laboratory Manual in Food Preservation.* AVI Publishing Co., Westport, CT.

Fish, B.P. 1958. Diffusion and thermodynamics of water in potato starch gel. In *Fundamental Aspects of the Dehydration of Foodstuffs*. Soc. Chem. Ind. (London) pp.143-157.

Gane, R. 1943. The activity of water in dried foodstuffs; water content as a function of humidity and temperature. In *Dehydration*, Sec. X, Part 1, U.K. Progress Report, D.S.I.R., Ministry of Food, London, U.K.

Gane, R. 1950. The water relation of some fruits, vegetables and plant products. *J. Soc. Food Agric.* 1:42-46.

Harkins, D. 1952. *The Physical Chemistry of Surface Films*. Van Nostrand Reinhold, New York.

Hayakawa, K., Nonino, C., and Succar, J. 1983. Two dimensional heat conduction in foods undergoing freezing: predicting freezing times of rectangular or finitely-cylindrical foods. *J. Food Sci.* 48:1841-1848.

Heldman, D.R., and Singh, R.P. 1981. *Food Process Engineering*. AVI Publishing Co., Westport, CT.

Hohner, G.A., and Heldman, D.R. 1970. Computer simulation of freezing rates in foods. Presented at 30th Annual Meeting of the *Institute of Food Technologists*. May 24-27, San Francisco.

Jason, A.C. 1958. A study of evaporation and diffusion processes in the drying of fish muscle. In *Fundamental Aspects of the Dehydration of Foodstuffs*. Soc. Chem. Ind. (London) pp.103-135.

Karel, M., Fennema, O.R., and Lund, D.B. 1975. *Principles of Food Science*, part II, Physical Principles of Food Preservation. Marcel Dekker, New York.

Kreith, F., and Black, W. Z. 1980. *Basic Heat Transfer*. Harper & Row, New York.

Mohsenin, N.N. 1980. *Thermal Properties of Foods and other Agricultural Materials*. Gordon and Breach, New York.

Moody, L.F. 1944. Friction factors for pipe flow. Trans. American Society of Mechanical Engineers. 66:671.

Murakami, E.G., Choi, Y., and Okos, M.R. 1985. An interactive computer program for predicting thermal properties. ASAE Paper No. 85-6511, American Society of Agricultural Engineers, St. Joseph, MI.

Nagaoka, J.S., Takagi, S., and Hotani, S. 1955. Experiments on the freezing of fish in an air-blast freezer. *Proc. 9th Inter. Congr. Refrig.* (Paris) 2,4.

Notter, G.K., Taylor, D.H., and Downes, N.J. 1959. Orange juice powder: factors affecting storage stability. *Food Technol.* 13:113-118.

Plank, R.Z. 1913. Ges. Kalte-Ind. 20,109 (in German). Cited by A.J. Ede. 1949. The calculation of the freezing and thawing of foodstuffs. *Mod. Refrig.* 52:52.

Shaw, D.J. 1980. *Introduction to Colloid and Surface Chemistry*. Butterworths, Oxford, U.K.

Siebel, J.E. 1892. Specific heat of various products. *Ice Refrig.* 2:256.

Stumbo, C.R. 1973. *Thermobacteriology in Food Processing*. Academic Press, New York.

Van Arsdel, W.B., Copley, M.J., and Morgen, A.I. 1973. *Food Dehydration*, Vol. 1. AVI Publishing Co., Westport, CT.

Van Wazer, J.R., Lyons, J.W., Kim, K.Y., and Colwell, R.E. 1963. *Viscosity and Flow Measurement*. Interscience Publishers, New York.

Van Wylen, G.J., and Sonntag, R.E. 1976. *Fundamentals of Classical Thermodynamics*. Wiley, New York.

Wark, K. 1983. *Thermodynamics*. McGraw Hill, New York.

Welty, J.R., Wicks, C.E., and Wilson, R.E. 1984. *Fundamentals of Momentum, Heat and Mass Transfer*. Wiley, New York.

Appendix F--Further Readings

Batty, J.C. and Folkman, S.L. 1983. *Food Engineering Fundamentals.* Wiley, New York.

Blakebrough, N. (ed.). 1968. *Biochemical and Biological Engineering Science.* Vol. 2. Academic Press, London, U.K.

Brennan, J.G., Butters, J.R., Cowell, N.D. and Lilly, A.E.V. 1976. *Food Engineering Operations.* Applied Science, London, U.K.

Charm, S.E. 1963. *Fundamentals of Food Engineering.* AVI Publishing Co., Westport, CT.

Earle, R.L. 1983. *Unit Operations in Food Processing.* Pergamon Press, Elmsford, N.Y.

Farrall, A.W. 1973. *Engineering for Dairy and Food Products.* Robert E. Krieger Publishing Co., Huntington, N.Y.

Farrall, A.W. 1976. *Food Engineering Systems. Vol. I-- Operations.* AVI Publishing Co., Westport, CT.

Farrall, A.W. 1979. *Food Engineering Systems. Vol. II-- Utilities.* AVI Publishing Co., Westport, CT.

Geankoplis, C.J. 1978. *Transport Process and Unit Operations.* Allyn and Bacon, Boston, MA.

Hall, C.W., Farrall, A.W. and Rippen, A.L. 1986. *Encyclopedia of Food Engineering.* AVI Publishing Co., Westport, CT.

Harper, W.J. and Hall, C.W. 1976. *Dairy Technology and Engineering.* AVI Publishing Co., Westport, CT.

Harper, J.C. 1976. *Elements of Food Engineering.* AVI Publishing Co., Westport, CT.

Heldman, D.R. and Singh, R.P. 1981. *Food Processing Engineering.* AVI Publishing Co., Westport, CT.

Henderson, S.M. and Perry, R.L. 1979. *Agricultural Process Engineering.* AVI Publishing Co., Westport, CT.

Jackson, A.T. and Lamb, J. 1981. *Calculations in Food and Chemical Engineering.* Macmillan Press, London, U.K.

Kessler, H.G. 1981. *Food Engineering and Dairy Technology.* Verlag A. Kessler, Freising, Germany.

Leniger, H.A. and Beverloo, W.A. 1975. *Food Process Engineering*. D. Reidel Publishing Co., Dordrecht, Holland.

Loncin, M. and Merson, R.L. 1979. *Food Engineering, Principles and Selected Applications*. Academic Press, New York.

Merkel, J.A. 1983. *Basic Engineering Principles*. AVI Publishing Co., Westport, CT.

Rao, M.A. and Rizvi, S.S.H. 1986. *Engineering Properties of Foods*. Marcel-Dekker, New York.

Singh, R.P and Heldman, D.R.. 1984. *Introduction to Food Engineering*. Academic Press, New York.

Toledo, R.T. 1980. *Fundamentals of Food Process Engineering*. AVI Publishing Co., Westport, CT.

Index